Workbook

Addition

A Direct Instruction Program

Siegfried Engelmann • Doug Carnine

Columbus, OH

The McGraw·Hill Companies

Cover and title page photo credits:
©Jose Luis Banus-March/Taxi/Getty Images, Inc.

SRAonline.com

SRA

Send all inquiries to:
SRA/McGraw-Hill
4400 Easton Commons
Columbus, OH 43219

Printed in the United States of America.

ISBN 0-07-602458-X

11 12 13 RHR 13 12

The McGraw·Hill Companies

Addition Placement Test

A

$$\begin{array}{r} 6 \\ +1 \\ \hline \end{array} \qquad \begin{array}{r} 8 \\ +1 \\ \hline \end{array} \qquad \begin{array}{r} 5 \\ +3 \\ \hline \end{array} \qquad \begin{array}{r} 3 \\ +0 \\ \hline \end{array} \qquad \begin{array}{r} 5 \\ +2 \\ \hline \end{array} \qquad \begin{array}{r} 7 \\ +1 \\ \hline \end{array}$$

$$\begin{array}{r} 5 \\ +4 \\ \hline \end{array} \qquad \begin{array}{r} 9 \\ +1 \\ \hline \end{array} \qquad \begin{array}{r} 8 \\ +0 \\ \hline \end{array} \qquad \begin{array}{r} 6 \\ +1 \\ \hline \end{array} \qquad \begin{array}{r} 5 \\ +5 \\ \hline \end{array} \qquad \begin{array}{r} 3 \\ +1 \\ \hline \end{array}$$

B

$$\begin{array}{r} 513 \\ 131 \\ +114 \\ \hline \end{array} \qquad \begin{array}{r} 23 \\ 32 \\ 50 \\ +21 \\ \hline \end{array} \qquad \begin{array}{r} 110 \\ 324 \\ 150 \\ +112 \\ \hline \end{array} \qquad \begin{array}{r} 31 \\ 22 \\ 52 \\ +41 \\ \hline \end{array} \qquad \begin{array}{r} 30 \\ 23 \\ 21 \\ +\ \ 2 \\ \hline \end{array} \qquad \begin{array}{r} 31 \\ 114 \\ 210 \\ +\ \ 11 \\ \hline \end{array}$$

C

$$\begin{array}{r} 1393 \\ 616 \\ +2482 \\ \hline \end{array} \qquad \begin{array}{r} 262 \\ 666 \\ +570 \\ \hline \end{array} \qquad \begin{array}{r} 169 \\ 377 \\ 22 \\ +\ \ 50 \\ \hline \end{array} \qquad \begin{array}{r} 1519 \\ 664 \\ +1612 \\ \hline \end{array} \qquad \begin{array}{r} 874 \\ 128 \\ 7 \\ +153 \\ \hline \end{array}$$

Part C continues on the next page.

275 men chop wood. 86 students work math problems. 42 girls chop wood. 36 men chop onions. 181 boys chop wood. Add the people who chop wood.

The store sells 91 skirts. It sells 59 beds. It sells 97 hats. It sells 11 coats. It sells 11 shirts. It fixes 23 shoes. Add the pieces of clothing the store sells.

Elaine sharpens 451 crayons. She uses 81 pencils. She sharpens 270 knives. She sharpens 161 scissors. Doug sharpens 215 pencils. Elaine sharpens 116 pencils. How many objects does Elaine sharpen?

There are 8 boys who help clean tables. There are 25 girls who play hockey. There are 25 boys who play hockey. There are 18 girls who play basketball. Add the children in our school who play sports.

Jill studied mathematics for 90 minutes. Tim studied spelling for 50 minutes. Jill took a nap for 20 minutes. Jill studied spelling for 50 minutes. Jill studied science for 200 minutes. How many minutes did Jill study?

Lesson 1

Facts + Bonus = TOTAL

1

9 + 1	6 + 1	2 + 0	5 + 1
0 + 0	1 + 1	3 + 0	6 + 0
7 + 1	5 + 0	9 + 0	8 + 1
7 + 0	2 + 1	8 + 0	4 + 1

2

A 264 **B** 538 **C** 490 **D** 751 **E** 892 **F** 376

3

6 3 8 7 2 9 1 4

Lesson 2

Facts + Problems + Bonus = TOTAL

1

	A	B	C	D	E
	65	23	35	41	57
	+ 24	+ 44	+ 57	+ 47	+ 1
	89	67	92	88	58

	F	G	H	I	J
	35	24	89	46	35
	+ 60	+ 1	+ 1	+ 0	+ 1
	95	25	90	46	36

2

2	7	3	9	6	5	1	10
+ 1	+ 0	+ 1	+ 1	+ 0	+ 1	+ 0	+ 1

8	2	7	20	8	3	4	2
+ 0	+ 0	+ 1	+ 0	+ 1	+ 0	+ 0	+ 0

3

A 7 { 6 — 6 + 1 = 7; 1 — 1 + 6 = 7

B 4 { 3 ______; 1 ______

C 5 { 4 ______; 1 ______

D 10 { 9 ______; 1 ______

4

A 279 **B** 146 **C** 657 **D** 328 **E** 864

5

6 3 7 5 9 2 10 4

Lesson 3

Facts + Problems + Bonus = TOTAL

1

A
$$\begin{array}{r} 56 \\ +41 \\ \hline 97 \end{array}$$

B
$$\begin{array}{r} 38 \\ +\ \ 4 \\ \hline 42 \end{array}$$

C
$$\begin{array}{r} 89 \\ +\ \ 1 \\ \hline 90 \end{array}$$

D
$$\begin{array}{r} 46 \\ +\ \ 4 \\ \hline 50 \end{array}$$

E
$$\begin{array}{r} 87 \\ +\ \ 1 \\ \hline 88 \end{array}$$

F
$$\begin{array}{r} 46 \\ +\ \ 0 \\ \hline 46 \end{array}$$

G
$$\begin{array}{r} 87 \\ +11 \\ \hline 98 \end{array}$$

H
$$\begin{array}{r} 58 \\ +\ \ 1 \\ \hline 59 \end{array}$$

I
$$\begin{array}{r} 53 \\ +\ \ 0 \\ \hline 53 \end{array}$$

J
$$\begin{array}{r} 43 \\ +24 \\ \hline 67 \end{array}$$

2

$$\begin{array}{r} 5 \\ +0 \\ \hline \end{array} \quad \begin{array}{r} 8 \\ +1 \\ \hline \end{array} \quad \begin{array}{r} 4 \\ +1 \\ \hline \end{array} \quad \begin{array}{r} 8 \\ +0 \\ \hline \end{array} \quad \begin{array}{r} 3 \\ +1 \\ \hline \end{array} \quad \begin{array}{r} 10 \\ +\ \ 0 \\ \hline \end{array} \quad \begin{array}{r} 1 \\ +1 \\ \hline \end{array} \quad \begin{array}{r} 6 \\ +1 \\ \hline \end{array}$$

$$\begin{array}{r} 9 \\ +1 \\ \hline \end{array} \quad \begin{array}{r} 0 \\ +0 \\ \hline \end{array} \quad \begin{array}{r} 7 \\ +0 \\ \hline \end{array} \quad \begin{array}{r} 5 \\ +1 \\ \hline \end{array} \quad \begin{array}{r} 6 \\ +0 \\ \hline \end{array} \quad \begin{array}{r} 0 \\ +1 \\ \hline \end{array} \quad \begin{array}{r} 4 \\ +0 \\ \hline \end{array} \quad \begin{array}{r} 9 \\ +0 \\ \hline \end{array}$$

3

A
[5] { 4 ____________ / 1 ____________

B
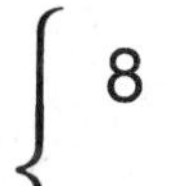
[9] { 8 ____________ / 1 ____________

C
[3] { 2 ____________ / 1 ____________

D
[7] { 6 ____________ / 1 ____________

4

A 425 B 738 C 916 D 437 E 514

5

Add 1 to each number.

5	18	14	9	13	6	12	7

Lesson 4

Facts	+	Problems	+	Bonus	=	TOTAL

1

A

6 { 6 ______ / 0 ______ }

B

4 { 3 ______ / 1 ______ }

C

11 { 10 ______ / 1 ______ }

D

3 { 3 ______ / 0 ______ }

E

4 { 4 ______ / 0 ______ }

F

8 { 7 ______ / 1 ______ }

2

A 356 B 297 C 424 D 932 E 358 F 672

3

5	8	9	10	0	4	8	3
+ 0	+ 1	+ 0	+ 0	+ 0	+ 1	+ 0	+ 1

1	4	2	5	4	0	8	5
+ 0	+ 1	+ 0	+ 1	+ 0	+ 1	+ 1	+ 0

4

A	B	C	D	E	F
43	53	34	57	42	46
+ 25	+ 24	+ 21	+ 32	+ 16	+ 15
68	77	55	89	58	61

5

Add 1 to each number.

8	12	3	16	14	10	5	17

Lesson 5

Facts + Bonus = TOTAL

1

A 9 ______ 1 ______

B 4 ______ 0 ______

C 3 ______ 0 ______

D 8 ______ 1 ______

2

A 406 **B** 308 **C** 728 **D** 507 **E** 346 **F** 205

3

$10 + 0$ $3 + 1$ $18 + 0$ $9 + 1$

$17 + 1$ $15 + 0$ $8 + 1$ $7 + 1$

$4 + 1$ $12 + 0$ $18 + 1$ $9 + 1$

$11 + 0$ $7 + 1$ $3 + 1$ $9 + 1$

Lesson 6

Facts	+	Problems	+	Bonus	=	TOTAL

1

A
6 ____________
0 ____________

B
9 ____________
1 ____________

C
5 ____________
1 ____________

D
5 ____________
0 ____________

E
3 ____________
1 ____________

F
3 ____________
0 ____________

2

$5 + 2$ $5 + 1$ $5 + 3$ $5 + 2$ $5 + 3$ $5 + 1$ $5 + 3$ $5 + 2$

3

A 508 B 326 C 409 D 583 E 704 F 542

4

$7 + 1$ $3 + 1$ $0 + 1$ $13 + 1$ $7 + 0$ $14 + 0$ $9 + 1$ $17 + 1$

$10 + 1$ $0 + 4$ $5 + 0$ $16 + 1$ $1 + 1$ $21 + 0$ $18 + 0$ $9 + 0$

5

Add 1 to each number.

8 13 2 27 13 5 14 18

Lesson 7

Test + Facts + Problems + Bonus = TOTAL

1

5	5	5	5	5	5	5	5
+ 4	+ 2	+ 4	+ 3	+ 2	+ 4	+ 3	+ 4

2

A 824 + 53 =

$$\begin{array}{r} 8\ 2\ 4 \\ +\ \_\ 5\ 3 \\ \hline \end{array}$$

B 621 + 58 =

C 302 + 4 =

D 521 + 19 =

E 53 + 6 =

F 206 + 23 =

3

A
$$\begin{array}{r} 51 \\ +32 \\ \hline 83 \end{array}$$

B
$$\begin{array}{r} 48 \\ +21 \\ \hline 69 \end{array}$$

C
$$\begin{array}{r} 93 \\ +\ \ 4 \\ \hline 97 \end{array}$$

D
$$\begin{array}{r} 53 \\ +\ \ 5 \\ \hline 58 \end{array}$$

E
$$\begin{array}{r} 73 \\ +\ \ 2 \\ \hline 75 \end{array}$$

4

Add 1 to each number.

14	8	25	17	9	12	3

5

A	306	425	814	730	502
B	211	908	460	322	177
C	690	843	364	505	491
D	121	615	709	410	852
E	326	180	978	732	101
F	881	562	453	207	320

6

15	8	1	16	1	18	11	17
+ 1	+ 0	+ 3	+ 1	+ 5	+ 1	+ 0	+ 1

0	1	5	0	19	1	0	6
+ 9	+ 2	+ 1	+ 3	+ 1	+ 8	+ 4	+ 1

7

Write the big number in the box.
Then write the two facts for each number family.

A ☐ { 5, 0 } ____________________ ____________________

B ☐ { 8, 1 } ____________________ ____________________

C ☐ { 4, 1 } ____________________ ____________________

D ☐ { 4, 0 } ____________________ ____________________

E ☐ { 2, 1 } ____________________ ____________________

F ☐ { 2, 0 } ____________________ ____________________

Lesson 8

Facts + Problems + Bonus = TOTAL

1

5	5	5	5	5	5	5	5
+ 5	+ 4	+ 5	+ 3	+ 2	+ 5	+ 4	+ 5

2

A 219 + 64 =

$$\begin{array}{r} 2\ 1\ 9 \\ +\ \ 6\ 4 \\ \hline \end{array}$$

B 851 + 62 =

C 504 + 78 =

D 520 + 6 =

E 224 + 401 =

F 523 + 4 =

3

14	1	8	1	16	0	1	17
+ 0	+ 9	+ 0	+ 3	+ 1	+ 7	+ 5	+ 1

0	15	20	1	7	0	1	21
+ 9	+ 1	+ 0	+ 8	+ 1	+ 4	+ 7	+ 0

4

Add 1 to each number.

14	8	19	3	12	23	5

5

	A	B	C	D	E
	153	321	426	308	504
+	424	34	31	161	23
	577	355	457	469	527

Lesson 9

Facts	+	Problems	+	Bonus	=	TOTAL

1

5	5	5	5	5	5	5	5
+ 5	+ 3	+ 5	+ 2	+ 4	+ 3	+ 5	+ 4

2

3	1	3	2	1	1	2	3
+ 2	+ 2	+ 2	+ 2	+ 2	+ 2	+ 2	+ 2

3

A 311 + 68 =

__ __ __
+ __ __ __

B 85 + 70 =

__ __ __
+ __ __ __

C 603 + 172 =

__ __ __
+ __ __ __

D 251 + 3 =

__ __ __
+ __ __ __

4

A	B	C	D	E
431	54	381	415	617
+ 15	+ 10	+ 5	+ 20	+ 50

5

1	0	1	1	20	0	1	12
+ 8	+ 8	+ 3	+ 10	+ 1	+ 9	+ 4	+ 0

10	15	1	15	0	27	12	1
+ 0	+ 1	+ 9	+ 0	+ 3	+ 1	+ 1	+ 4

6

Add 1 to each number.

17	3	14	8	24	12	15

Lesson 10

Facts + Bonus = TOTAL

1

5 + 3	5 + 5	5 + 2	5 + 4	5 + 5	5 + 3	5 + 2	5 + 5
5 + 3	5 + 1	5 + 0	5 + 2	5 + 5	5 + 4	5 + 3	5 + 4

2

4 + 2	1 + 2	3 + 2	4 + 2	3 + 2	4 + 2	3 + 2	1 + 2

3

A	B	C	D	E
1	1	4	5	4
4	3	1	2	1
2	1	3	1	0
+ 1	+ 2	+ 1	+ 1	+ 2

4

A 367 + 4 =

B 809 + 20 =

C 802 + 14 =

D 80 + 46 =

5

A	B	C	D	E
402	35	41	31	320
+ 31	+ 210	+ 13	+ 405	+ 10

6

8 + 1	15 + 0	3 + 1	0 + 10	1 + 9	16 + 0	4 + 1	1 + 3
15 + 1	0 + 6	0 + 7	3 + 1	12 + 0	1 + 5	0 + 3	14 + 1

7

Write the big number in the box.
Then write the two facts for each number family.

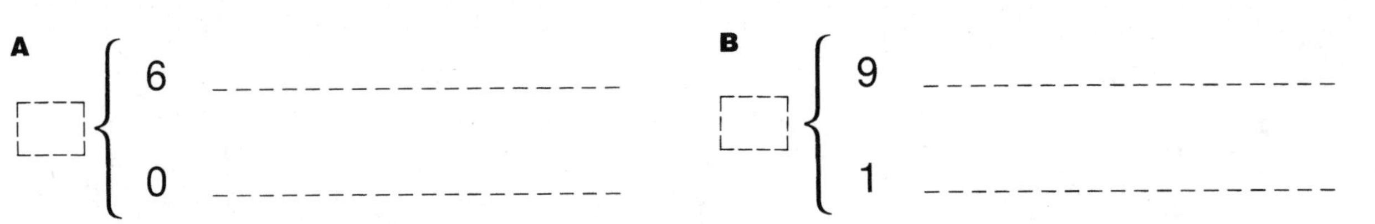

C ☐ { 5, 1

D ☐ { 5, 0

E ☐ { 3, 1

F ☐ { 3, 0

Lesson 11

Test	+	Facts	+	Problems	+	Bonus	=	TOTAL

1

5 + 3	5 + 2	5 + 4	5 + 5	5 + 0	5 + 5	5 + 3	5 + 4
5 + 4	5 + 5	5 + 1	5 + 3	5 + 0	5 + 4	5 + 3	5 + 5

2

5 + 2	2 + 2	4 + 2	3 + 2	1 + 2	4 + 2	3 + 2	5 + 2

3

A	B	C	D	E
3 1 2 + 1	5 2 1 + 1	1 4 3 + 1	1 3 1 + 3	1 2 1 + 1

4

A 305 + 21 = ______ **B** 420 + 8 = ______ **C** 350 + 104 = ______

5

3 + 1	1 + 10	0 + 8	4 + 1	19 + 0	6 + 1	11 + 1	1 + 3
10 + 0	0 + 5	19 + 1	1 + 0	3 + 1	18 + 0	14 + 1	1 + 2

Write the big number in the box.
Then write the two facts for each number family.

A

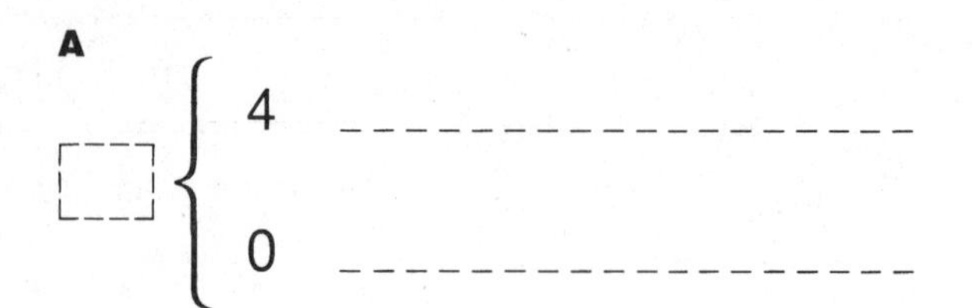

B

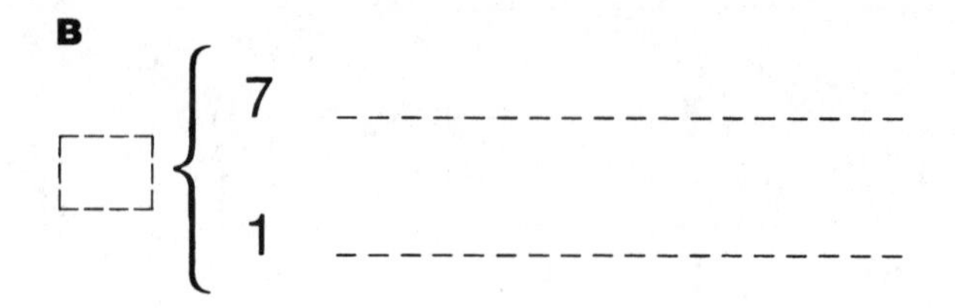

C

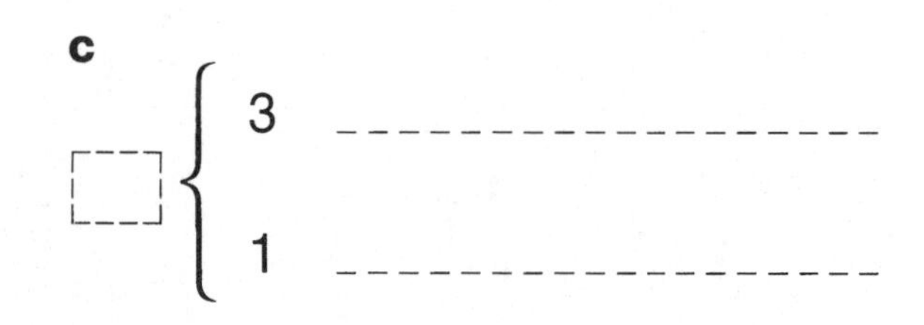

D

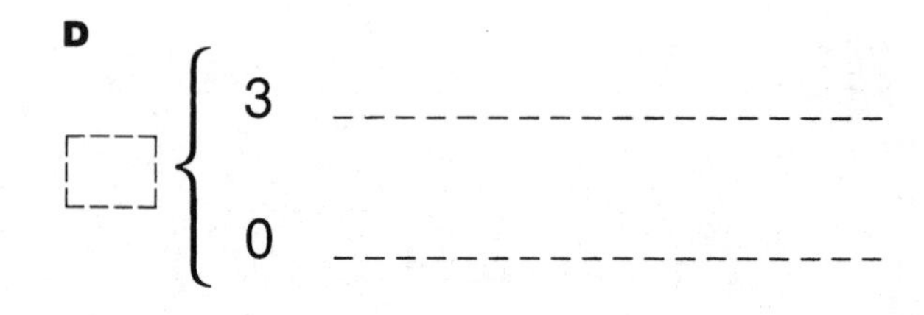

E

☐ { 1 ____________ / 1 ____________

F

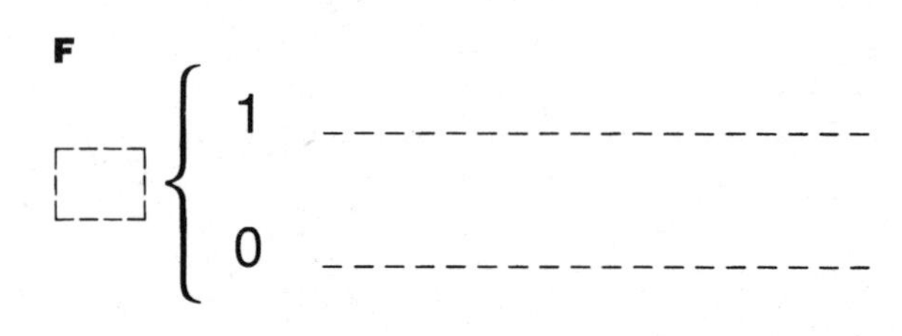

7

A	B	C	D	E
403 + 21	85 + 10	561 + 104	381 + 2	45 + 101

8

Add 1 to each number.

14	9	27	12	17	15	6
☐	☐	☐	☐	☐	☐	☐

Lesson 12

Facts	+	Bonus	=	TOTAL

1

3	5	2	1	4	2	5	3
+ 2	+ 2	+ 2	+ 2	+ 2	+ 2	+ 2	+ 2

2

A [ ] { 5 ______ 2 ______

B [ ] { 5 ______ 4 ______

C [ ] { 5 ______ 1 ______

D [ ] { 5 ______ 3 ______

3

5	0	1	1	5	5	8	5
+ 3	+ 9	+ 8	+ 8	+ 2	+ 5	+ 1	+ 4

1	5	1	0	0	5	5	5
+ 6	+ 5	+ 5	+ 12	+ 8	+ 1	+ 4	+ 3

4

A	B	C	D	E
1	1	1	1	4
3	4	5	1	1
1	2	1	3	3
+ 4	+ 1	+ 1	+ 2	+ 1

5

A 402 + 31 = ______ **B** 321 + 5 = ______ **C** 134 + 401 = ______

Lesson 13

Facts + Problems + Bonus = TOTAL

1

1	5	0	3	4	2	5	1
+ 2	+ 2	+ 2	+ 2	+ 2	+ 2	+ 2	+ 2

2	0	5	4	3	2	4	3
+ 2	+ 2	+ 2	+ 2	+ 2	+ 2	+ 2	+ 2

2

5	1	5	3	0	5	10	5
+ 4	+ 2	+ 2	+ 0	+ 15	+ 3	+ 1	+ 5

9	15	16	5	5	0	5	0
+ 1	+ 1	+ 0	+ 5	+ 3	+ 10	+ 2	+ 8

3

A [ ] { 5 ________ / 3 ________

B [ ] { 5 ________ / 2 ________

C [ ] { 5 ________ / 0 ________

D [ ] { 5 ________ / 4 ________

E [ ] { 5 ________ / 5 ________

F [ ] { 5 ________ / 1 ________

4

A 305 + 23 = ______

B 251 + 6 = ______

Part 4 continues on the next page.

C 520 + 300 = ______ D 53 + 20 = ______

5

A	B	C	D	E	F
2	0	5	1	2	5
1	4	3	4	3	5
2	1	1	5	4	1
+ 4	+ 5	+ 1	+ 1	+ 1	+ 1

6

A ___ 9 6

B 7 4 2

C ___ ___ ___

D ___ ___ ___

E ___ ___ ___

F ___ ___ ___

G ___ ___ ___

Lesson 14

Facts + Problems + Bonus = TOTAL

1

A 4 2 6

B ___ ___ ___

C ___ ___ ___

D ___ ___ ___

E ___ ___ ___

F ___ ___ ___

2

A $352 + 40 =$ ______

B $408 + 100 =$ ______

C $500 + 60 =$ ______

D $305 + 4 =$ ______

3

A	B	C	D	E	F
5	3	3	1	1	5
2	1	2	9	3	4
1	1	5	0	1	1
+ 0	+ 5	+ 1	+ 1	+ 5	+ 1

4

4	2	0	5	1	3	5	2
+2	+2	+2	+2	+2	+2	+2	+2

3	0	4	2	5	4	3	5
+ 2	+ 2	+ 2	+ 2	+ 2	+ 2	+ 2	+ 2

5

4 + 5	2 + 5	1 + 5	3 + 5	0 + 5	5 + 5	2 + 5	4 + 5
3 + 5	5 + 5	2 + 5	4 + 5	1 + 5	3 + 5	0 + 5	1 + 5

6

1 + 8	5 + 3	0 + 9	11 + 0	5 + 4	0 + 5	5 + 2	1 + 5
16 + 1	5 + 2	12 + 1	0 + 4	13 + 1	5 + 3	15 + 1	5 + 4

7

Write the big number in the box.
Then write the two facts for each number family.

A [ ] { 5, 3 } ____________________

B [ ] { 5, 2 } ____________________

C [ ] { 5, 0 } ____________________

D [ ] { 5, 4 } ____________________

E [ ] { 5, 5 } ____________________

F [ ] { 5, 1 } ____________________

Lesson 15

Facts + Problems + Bonus = TOTAL

1

2	3	1	4	0	5	2	4
+ 5	+ 5	+ 5	+ 5	+ 5	+ 5	+ 5	+ 5

3	5	1	4	0	3	2	5
+ 5	+ 5	+ 5	+ 5	+ 5	+ 5	+ 5	+ 5

2

4	5	12	2	3	3	0	5
+ 2	+ 3	+ 0	+ 2	+ 1	+ 2	+ 2	+ 2

5	3	5	17	1	14	4	3
+ 4	+ 2	+ 1	+ 0	+ 2	+ 1	+ 2	+ 2

3

A { 1 ________ 2 ________

B { 4 ________ 2 ________

C { 5 ________ 2 ________

D { 3 ________ 2 ________

E { 0 ________ 2 ________

F { 4 ________ 2 ________

4

A 408 + 10 = ______ B 507 + 300 = ______ C 158 + 41 = ______

D 365 + 3 = ______

5

A	B	C	D	E	F
21	30	12	14	13	10
33	23	12	31	41	13
31	11	32	10	31	32
+ 12	+ 13	+ 31	+ 21	+ 11	+ 20

6

A 4 2 8

B ___ ___ ___

C ___ ___ ___

D ___ ___ ___

E ___ ___ ___

F ___ ___ ___

Lesson 16

Test	+	Facts	+	Problems	+	Bonus	=	TOTAL

1

1 + 5	4 + 5	2 + 5	0 + 5	5 + 5	3 + 5	1 + 5	4 + 5
2 + 5	5 + 5	3 + 5	0 + 5	4 + 5	2 + 5	5 + 5	3 + 5

2

A { 3 ________ ; 2 ________ }

B { 1 ________ ; 2 ________ }

C { 5 ________ ; 2 ________ }

D { 4 ________ ; 2 ________ }

3

10 + 3	10 + 1	10 + 2	10 + 5	10 + 4	10 + 3	10 + 2	10 + 5

4

5 + 2	5 + 3	11 + 0	3 + 2	1 + 2	15 + 0	4 + 2	2 + 2
13 + 1	5 + 4	4 + 2	2 + 2	19 + 1	14 + 1	5 + 2	3 + 2
5 + 5	3 + 2	5 + 3	4 + 2	5 + 4	0 + 8	3 + 2	5 + 4

5

A 4 2 5

B ___ ___ ___

C ___ ___ ___

D ___ ___ ___

E ___ ___ ___

F ___ ___ ___

6

A	B	C	D	E	F
11	13	11	31	30	3
13	32	32	22	23	11
31	10	15	10	21	21
+ 14	+ 21	+ 11	+ 15	+ 11	+ 1

7

A $527 + 30 =$ ______

B $314 + 20 =$ ______

C $305 + 43 =$ ______

D $413 + 170 =$ ______

Lesson 17

Facts	+	Problems	+	Bonus	=	TOTAL

1

A	B	C	D	E	F
51	52	24	11	35	41
52	53	22	41	21	52
21	11	50	53	10	15
+ 12	+ 11	+ 11	+ 12	+ 2	+ 1
6	7				

2

A 2 0 7

B ___ ___ ___

C ___ ___ ___

D ___ ___ ___

E ___ ___ ___

F ___ ___ ___

3

A 403 + 25 = ______

B 525 + 3 = ______

C 501 + 35 = ______

D 235 + 4 = ______

4

10	10	10	10	10	10	10	10
+ 2	+ 5	+ 4	+ 1	+ 3	+ 4	+ 2	+ 5

5

2 + 4	2 + 1	2 + 3	2 + 2	2 + 5	2 + 3	2 + 0	2 + 4
2 + 3	2 + 5	2 + 4	2 + 2	2 + 4	2 + 5	2 + 3	2 + 2

6

1 + 5	14 + 1	3 + 5	4 + 2	4 + 5	10 + 1	2 + 5	5 + 5
3 + 2	4 + 5	0 + 5	5 + 4	3 + 5	13 + 1	5 + 3	2 + 5

7

A ☐ { 1 ______ 2 ______

B ☐ { 4 ______ 2 ______

C ☐ { 5 ______ 2 ______

D ☐ { 3 ______ 2 ______

E ☐ { 0 ______ 2 ______

F ☐ { 4 ______ 2 ______

Lesson 18

Facts	+	Problems	+	Bonus	=	TOTAL

1

10	10	10	10	10	10	10	10
+ 5	+ 2	+ 1	+ 4	+ 3	+ 2	+ 4	+ 5

2

A	B	C	D	E	F
41	84	52	52	91	13
12	22	42	1	11	42
55	31	11	+ 15	10	42
+ 41	+ 21	+ 10		+ 12	+ 1
9	8				

3

A 3 2 4

B ___ ___ ___

C ___ ___ ___

D ___ ___ ___

E ___ ___ ___

F ___ ___ ___

4

A 424 + 15 = ______

B 350 + 42 = ______

C 485 + 13 = ______

D 207 + 11 = ______

5

$$\begin{array}{r}2\\+1\\\hline\end{array}\quad\begin{array}{r}2\\+5\\\hline\end{array}\quad\begin{array}{r}2\\+4\\\hline\end{array}\quad\begin{array}{r}2\\+2\\\hline\end{array}\quad\begin{array}{r}2\\+3\\\hline\end{array}\quad\begin{array}{r}2\\+0\\\hline\end{array}\quad\begin{array}{r}2\\+5\\\hline\end{array}\quad\begin{array}{r}2\\+3\\\hline\end{array}$$

$$\begin{array}{r}2\\+4\\\hline\end{array}\quad\begin{array}{r}2\\+2\\\hline\end{array}\quad\begin{array}{r}2\\+5\\\hline\end{array}\quad\begin{array}{r}2\\+3\\\hline\end{array}\quad\begin{array}{r}2\\+1\\\hline\end{array}\quad\begin{array}{r}2\\+4\\\hline\end{array}\quad\begin{array}{r}2\\+2\\\hline\end{array}\quad\begin{array}{r}2\\+5\\\hline\end{array}$$

6

$$\begin{array}{r}5\\+2\\\hline\end{array}\quad\begin{array}{r}4\\+5\\\hline\end{array}\quad\begin{array}{r}1\\+5\\\hline\end{array}\quad\begin{array}{r}0\\+7\\\hline\end{array}\quad\begin{array}{r}5\\+5\\\hline\end{array}\quad\begin{array}{r}2\\+5\\\hline\end{array}\quad\begin{array}{r}3\\+2\\\hline\end{array}\quad\begin{array}{r}3\\+5\\\hline\end{array}$$

$$\begin{array}{r}0\\+4\\\hline\end{array}\quad\begin{array}{r}4\\+2\\\hline\end{array}\quad\begin{array}{r}3\\+5\\\hline\end{array}\quad\begin{array}{r}16\\+\ \ 1\\\hline\end{array}\quad\begin{array}{r}2\\+5\\\hline\end{array}\quad\begin{array}{r}5\\+5\\\hline\end{array}\quad\begin{array}{r}12\\+\ \ 1\\\hline\end{array}\quad\begin{array}{r}4\\+5\\\hline\end{array}$$

7

A [ ] { 6 ________ / 0 ________ }

B [ ] { 9 ________ / 1 ________ }

C [ ] { 5 ________ / 1 ________ }

D [ ] { 5 ________ / 0 ________ }

E [ ] { 3 ________ / 1 ________ }

F [ ] { 3 ________ / 0 ________ }

Lesson 19

Facts + Problems + Bonus = TOTAL

1

2	2	2	2	2	2	2	2
+ 4	+ 0	+ 1	+ 3	+ 5	+ 2	+ 0	+ 4

2	2	2	2	2	2	2	2
+ 3	+ 5	+ 2	+ 4	+ 1	+ 3	+ 4	+ 5

2

10	10	10	10	10	10	10	10
+ 8	+ 6	+ 9	+ 7	+ 8	+ 3	+ 5	+ 7

3

A

5	5	5	5	5	5	5	5
+ 4	+ 14	+ 24	+ 34	+ 44	+ 54	+ 64	+ 74

B

3	3	3	3	3	3	3	3
+ 5	+ 15	+ 25	+ 35	+ 45	+ 55	+ 65	+ 75

4

2	1	7	4	1	13	5	0
+ 5	+ 9	+ 0	+ 5	+ 8	+ 1	+ 5	+ 6

3	1	14	5	1	9	2	13
+ 2	+ 5	+ 1	+ 4	+ 2	+ 0	+ 2	+ 1

5	2	1	4	0	1	3	2
+ 0	+ 3	+ 8	+ 5	+ 8	+ 3	+ 5	+ 4

5

A Add the plants. **B** Add the CDs. **C** Add the clocks.

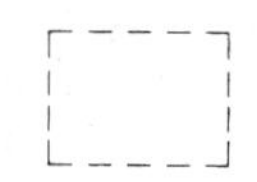
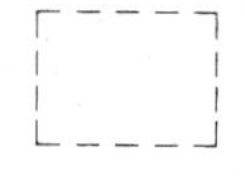

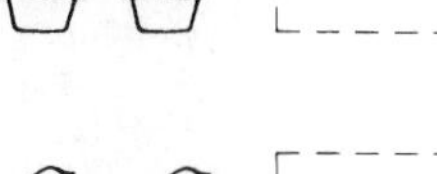

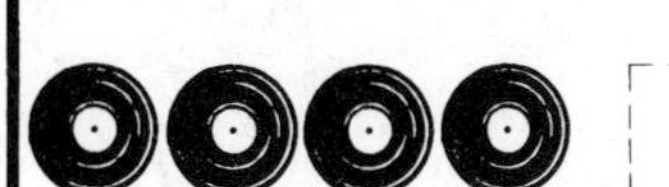

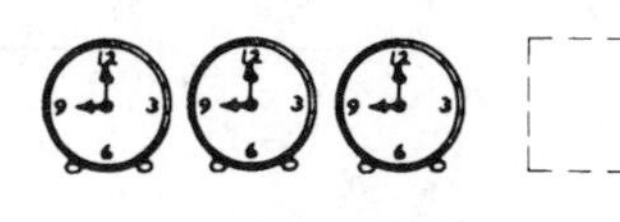

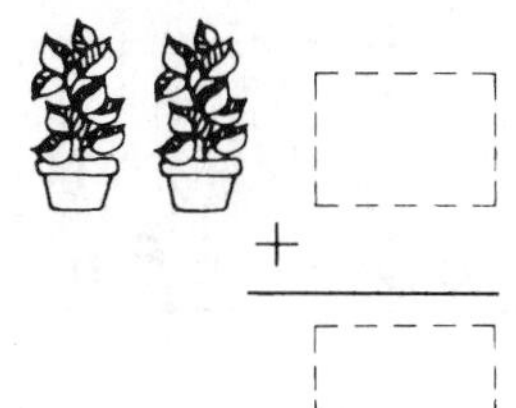
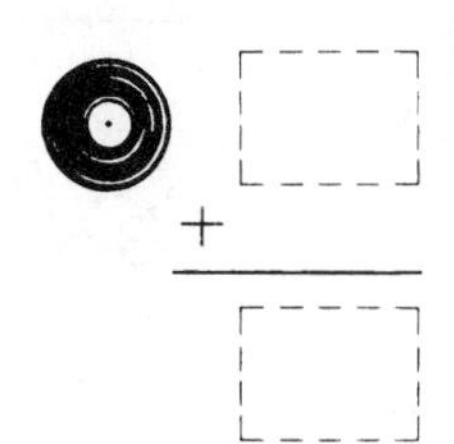
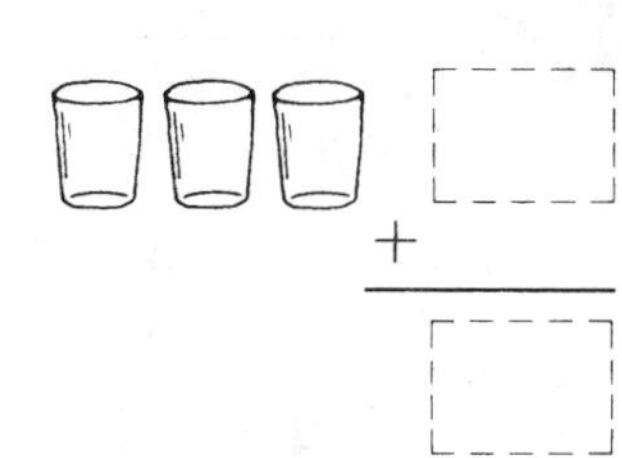

6

A	B	C	D	E	F
72	2	42	23	23	92
55	83	11	22	34	1
31	30	52	+ 11	1	12
+ 20	+ 42	+ 2		+ 1	+ 13
8	7				

7

A ___ ___ 8

B ___ ___ ___

C ___ ___ ___

D ___ ___ ___

E ___ ___ ___

F ___ ___ ___

8

A 106 + 22 = ______ **B** 302 + 5 = ______

Lesson 20

Facts	+	Problems	+	Bonus	=	TOTAL

1

10 + 7	10 + 10	10 + 8	10 + 9	10 + 6	10 + 8	10 + 9	10 + 3
10 + 6	10 + 1	10 + 7	10 + 5	10 + 7	10 + 10	10 + 9	10 + 8

2

A	4 + 5	4 + 15	4 + 25	4 + 35	4 + 45	4 + 55	4 + 65	4 + 75
B	3 + 5	3 + 15	3 + 25	3 + 35	3 + 45	3 + 55	3 + 65	3 + 75

3

2 + 3	4 + 5	2 + 4	12 + 1	2 + 5	5 + 5	2 + 5	3 + 5
14 + 1	4 + 2	5 + 0	2 + 3	3 + 1	2 + 5	2 + 4	0 + 8
5 + 4	2 + 5	0 + 7	5 + 5	9 + 1	5 + 3	2 + 4	1 + 8

A Add the glasses.

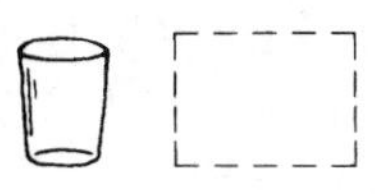

+

B Add the houses.

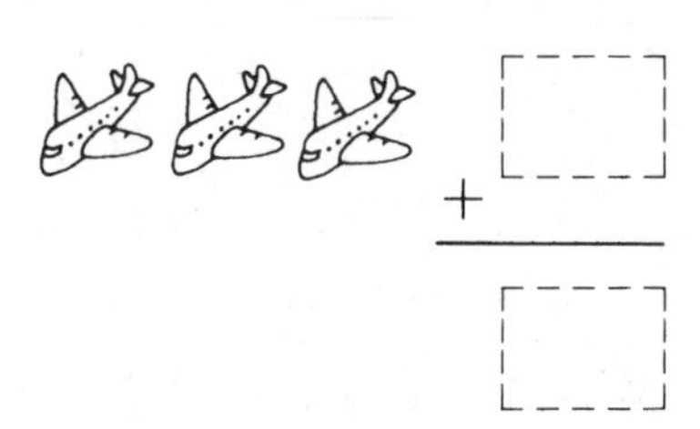

+

C Add the cups.

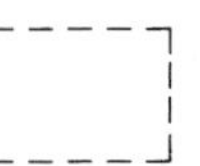

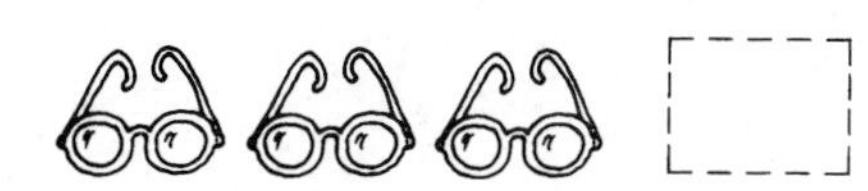

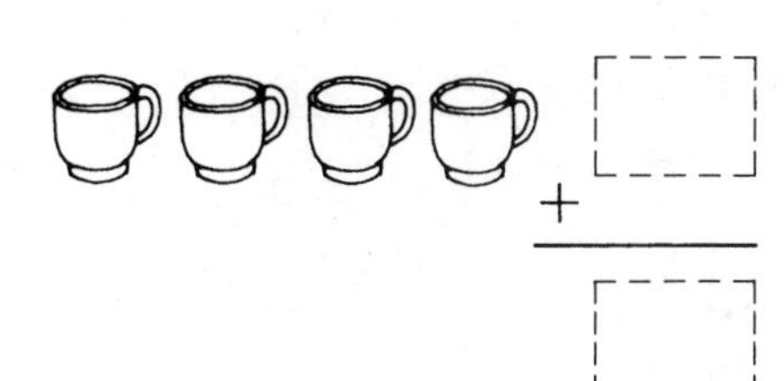

+

D Add the trees.

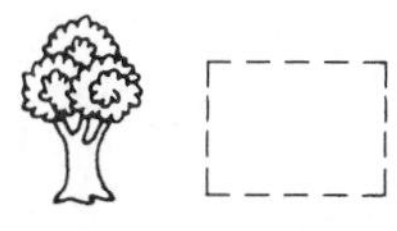

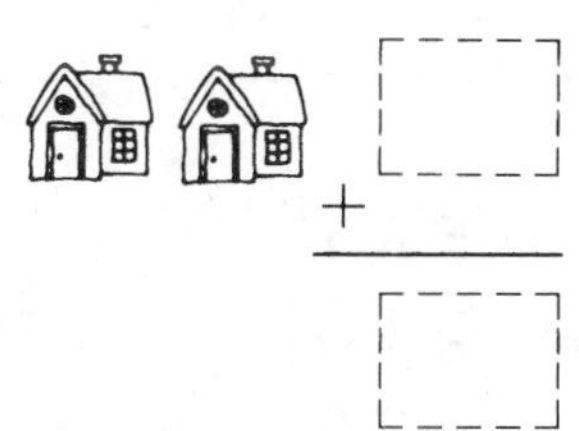

+

5

A

Add the number of shirts Sam sews.

1. Sam sews 5 shirts. _____
2. Sam sells 7 shirts. _____
3. Sam buys 3 jackets. _____
4. Sam sews 4 shirts. _____
5. Sam sews 1 shirt. + _____

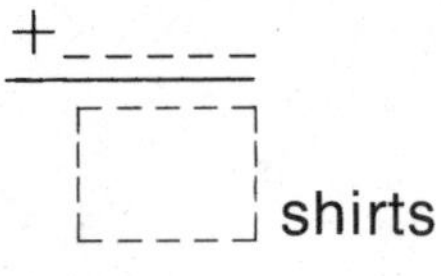
shirts

B

Add the number of eggs Carol eats.

1. Carol eats 1 egg. _____
2. Carol cooks 1 potato. _____
3. Carol eats 4 eggs. _____
4. Carol eats 2 eggs. _____
5. Carol cooks 3 eggs. + _____

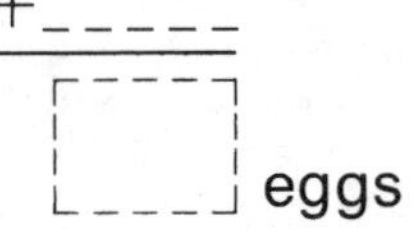
eggs

6

A

$$\begin{array}{r} 20 \\ 30 \\ 30 \\ +10 \\ \hline \end{array}$$

B

$$\begin{array}{r} 40 \\ 20 \\ 10 \\ +10 \\ \hline \end{array}$$

C

$$\begin{array}{r} 80 \\ 20 \\ 40 \\ +10 \\ \hline \end{array}$$

7

A ___ ___ 3

B ___ ___ ___

C ___ ___ ___

D ___ ___ ___

E ___ ___ ___

F ___ ___ ___

8

A

$$\begin{array}{r} 531 \\ 210 \\ 20 \\ +114 \\ \hline \end{array}$$

B

$$\begin{array}{r} 112 \\ 312 \\ 2 \\ +\ \ 51 \\ \hline \end{array}$$

C

$$\begin{array}{r} 321 \\ 220 \\ 211 \\ +\ \ 31 \\ \hline \end{array}$$

D

$$\begin{array}{r} 51 \\ 54 \\ 33 \\ +10 \\ \hline \end{array}$$

E

$$\begin{array}{r} 32 \\ 20 \\ 53 \\ +31 \\ \hline \end{array}$$

Lesson 21

Test	+	Facts	+	Problems	+	Bonus	=	TOTAL

1

$$\begin{array}{r} 10 \\ +\ 6 \\ \hline \end{array} \quad \begin{array}{r} 10 \\ +\ 9 \\ \hline \end{array} \quad \begin{array}{r} 10 \\ +\ 8 \\ \hline \end{array} \quad \begin{array}{r} 10 \\ +\ 2 \\ \hline \end{array} \quad \begin{array}{r} 10 \\ +\ 5 \\ \hline \end{array} \quad \begin{array}{r} 10 \\ +\ 7 \\ \hline \end{array} \quad \begin{array}{r} 10 \\ +\ 3 \\ \hline \end{array} \quad \begin{array}{r} 10 \\ +10 \\ \hline \end{array}$$

$$\begin{array}{r} 10 \\ +\ 9 \\ \hline \end{array} \quad \begin{array}{r} 10 \\ +\ 5 \\ \hline \end{array} \quad \begin{array}{r} 10 \\ +\ 8 \\ \hline \end{array} \quad \begin{array}{r} 10 \\ +\ 3 \\ \hline \end{array} \quad \begin{array}{r} 10 \\ +\ 7 \\ \hline \end{array} \quad \begin{array}{r} 10 \\ +\ 6 \\ \hline \end{array} \quad \begin{array}{r} 10 \\ +\ 2 \\ \hline \end{array} \quad \begin{array}{r} 10 \\ +\ 4 \\ \hline \end{array}$$

2

A

Add the number of homes that Jim decorates.

1. Jim buys 2 homes. _____
2. Jim drives by 8 homes. _____
3. Jim decorates 4 homes. _____
4. Jim decorates 1 home. _____
5. Jim visits 7 homes. + _____

[ ] homes

B

Find out how many hats the girl wears.

1. The girl wears 3 hats. _____
2. The girl wears 1 hat. _____
3. The girl wears 1 hat. _____
4. The girl wears 2 socks. _____
5. The girl wears 2 hats. + _____

[ ] hats

C

Find out how many cars Miss Blue Eagle sold.

1. Miss Blue Eagle sold 4 tires. _____
2. Miss Blue Eagle sold 5 cars. _____
3. Miss Blue Eagle sold 1 car. _____
4. Miss Blue Eagle saw 2 cars. _____
5. Miss Blue Eagle sold 1 car. + _____

[ ] cars

D

Add the number of shirts that Joe cleans.

1. Joe wears 1 shirt. _____
2. Joe cleans 6 shoes. _____
3. Joe cleans 5 shirts. _____
4. Joe cleans 4 shirts. _____
5. Joe buys 3 shirts. + _____

[ ] shirts

3

A

4	4	4	4	4	4	4	4
+ 2	+ 12	+ 22	+ 32	+ 42	+ 52	+ 62	+ 72

B

5	15	25	35	45	55	65	75
+ 3	+ 3	+ 3	+ 3	+ 3	+ 3	+ 3	+ 3

4

2	2	1	2	5	2	0
+ 4	+ 2	+ 5	+ 5	+ 4	+ 3	+ 8

3	4	2	4	5	2	5
+ 5	+ 5	+ 3	+ 2	+ 3	+ 4	+ 2

5

A	B	C	D
50	20	50	10
50	40	40	10
30	10	10	30
+ 10	+ 10	+ 10	+ 40

6

A ___ ___ 4

B ___ ___ ___

C ___ ___ ___

D ___ ___ ___

E ___ ___ ___

F ___ ___ ___

7

A	B	C	D	E
42	110	53	91	53
55	2	51	15	100
+ 11	551	+ 52	91	4
	+ 15		+ 10	+ 30

Lesson 22

Facts + Problems + Bonus = TOTAL

1

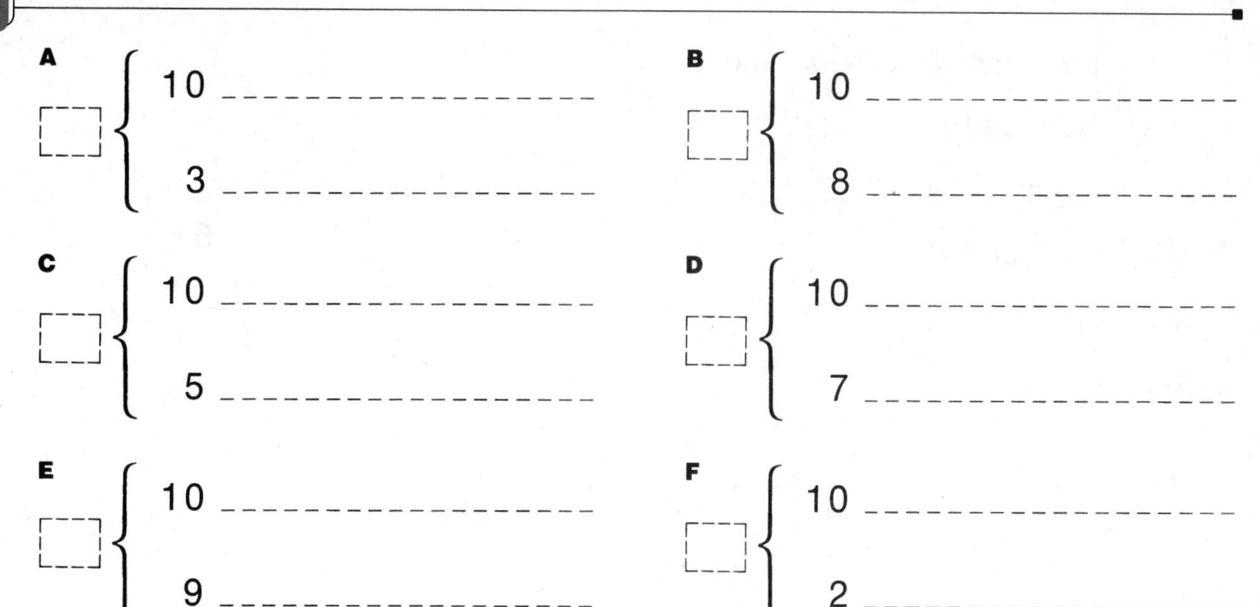

2

10 + 5	2 + 4	10 + 8	3 + 5	10 + 6	10 + 2	2 + 5	2 + 3
10 + 3	2 + 5	10 + 9	4 + 5	10 + 4	10 + 1	5 + 5	5 + 3

3

A

Add the pictures Flavio painted.

1. Flavio painted 2 pictures. ___
2. Flavio painted 3 pictures. ___
3. Flavio painted 1 picture. ___
4. Christina painted 2 pictures. ___
5. Flavio painted 2 pictures. + ___

[ ] ______________

Part 3 continues on the next page.

B

Find out how many bikes Mia fixed.

1. Mia fixed 5 bikes. ___
2. Mia bought 4 bikes. ___
3. Mia fixed 2 bikes. ___
4. Mia fixed 1 bike. ___
5. Mia fixed 1 bike. + ___

[ ] ____________

D

Add the birds that fly high.

1. 4 birds fly low. ___
2. 5 birds fly high. ___
3. 2 birds sing high. ___
4. 2 planes fly high. + ___

[ ] ______________

C

Add the people who dream.

1. 2 people dream. ___
2. 3 people dream. ___
3. 5 people run. ___
4. 3 people dream. ___
5. 2 dogs dream. + ___

[ ] ______________

4

A

$$\begin{array}{r} 3 \\ +5 \\ \hline \end{array} \quad \begin{array}{r} 3 \\ +15 \\ \hline \end{array} \quad \begin{array}{r} 3 \\ +25 \\ \hline \end{array} \quad \begin{array}{r} 3 \\ +35 \\ \hline \end{array} \quad \begin{array}{r} 3 \\ +45 \\ \hline \end{array} \quad \begin{array}{r} 3 \\ +55 \\ \hline \end{array} \quad \begin{array}{r} 3 \\ +65 \\ \hline \end{array} \quad \begin{array}{r} 3 \\ +75 \\ \hline \end{array}$$

B

$$\begin{array}{r} 2 \\ +4 \\ \hline \end{array} \quad \begin{array}{r} 12 \\ +4 \\ \hline \end{array} \quad \begin{array}{r} 22 \\ +4 \\ \hline \end{array} \quad \begin{array}{r} 32 \\ +4 \\ \hline \end{array} \quad \begin{array}{r} 42 \\ +4 \\ \hline \end{array} \quad \begin{array}{r} 52 \\ +4 \\ \hline \end{array} \quad \begin{array}{r} 62 \\ +4 \\ \hline \end{array} \quad \begin{array}{r} 72 \\ +4 \\ \hline \end{array}$$

5

A ___ ___ 8

B ___ ___ ___

C ___ ___ ___

D ___ ___ ___

E ___ ___ ___

F ___ ___ ___

6

A

```
  30
  50
  10
+ 10
----
```

B

```
  410
  534
    1
+  23
-----
```

C

```
  10
  40
  50
+ 60
----
```

7

A $125 + 32 =$ ______

```
  ___ ___ ___
+ ___ ___ ___
-------------
```

B $305 + 4 =$ ______

```
  ___ ___ ___
+ ___ ___ ___
-------------
```

Lesson 23

Facts	+	Problems	+	Bonus	=	TOTAL

1

A
10 ______
4 ______

B
10 ______
1 ______

C
10 ______
6 ______

D
10 ______
4 ______

E
10 ______
7 ______

F
10 ______
9 ______

2

2 + 4	4 + 5	10 + 3	2 + 5	10 + 9	3 + 5	2 + 3	10 + 5
10 + 4	4 + 2	10 + 2	0 + 7	10 + 6	3 + 5	10 + 8	0 + 6

3

5 + 6	5 + 5	5 + 7	5 + 5	5 + 7	5 + 6	5 + 5	5 + 7

4

A

5 + 4	5 + 14	5 + 24	5 + 34	5 + 44	5 + 54	5 + 64	5 + 74

B

4 + 2	14 + 2	24 + 2	34 + 2	44 + 2	54 + 2	64 + 2	74 + 2

A

Add the nails that Kim pounded.

1. Kim pounded 4 nails. ___
2. Al pounded 2 nails. ___
3. Kim pounded 2 nails. ___
4. Kim pounded 3 drums. ___
5. Kim pounded 1 nail. + ___

[ ] ________________

B

Find out how many pies Ann ate.

1. Ann ate 3 cakes. ___
2. Charles ate 2 pies. ___
3. Ann ate 5 pies. ___
4. Ann ate 5 pies. ___
5. Ann ate 1 pie. + ___

[ ] ________________

C

How many games did the children play?

1. The children played 3 games. ___
2. The parents played 7 games. ___
3. The children played 2 games. ___
4. The children played 5 games. ___
5. The children won 6 games. + ___

[ ] ________________

Part 5 continues on the next page.

D

Find out how many trees fell.

1. 5 trees fell. ___
2. 4 trees grew. ___
3. 2 trees fell. ___
4. 1 tree grew. ___
5. 6 leaves fell. + ___

[ ] ____

6

A ___ ___ 3

B ___ ___ ___

C ___ ___ ___

D ___ ___ ___

E ___ ___ ___

F ___ ___ ___

7

A
```
   503
     2
    20
+   52
------
```

B
```
  50
  50
  10
+ 10
----
```

C
```
  52
  51
  40
+ 11
----
```

Lesson 24

Facts + Problems + Bonus = TOTAL

1

$4 + 10$ | $6 + 10$ | $9 + 10$ | $2 + 10$ | $5 + 10$ | $1 + 10$ | $8 + 10$ | $7 + 10$

$3 + 10$ | $1 + 10$ | $7 + 10$ | $4 + 10$ | $6 + 10$ | $9 + 10$ | $2 + 10$ | $8 + 10$

2

$2 + 5$ | $10 + 7$ | $2 + 4$ | $10 + 5$ | $10 + 8$ | $2 + 3$ | $10 + 6$ | $0 + 4$

$10 + 1$ | $5 + 3$ | $10 + 4$ | $1 + 5$ | $5 + 2$ | $10 + 3$ | $10 + 9$ | $3 + 5$

3

A

$5 + 3$ | $5 + 13$ | $5 + 23$ | $5 + 33$ | $5 + 43$ | $5 + 53$ | $5 + 63$ | $5 + 73$

B

$4 + 5$ | $14 + 5$ | $24 + 5$ | $34 + 5$ | $44 + 5$ | $54 + 5$ | $64 + 5$ | $74 + 5$

4

$5 + 8$ | $5 + 6$ | $5 + 8$ | $5 + 7$ | $5 + 6$ | $5 + 8$ | $5 + 6$ | $5 + 7$

5

A

How many carrots did Bill chop?

1. Bill chopped 2 carrots. ___
2. Bill chopped 4 nuts. ___
3. Bill chopped 1 carrot. ___
4. Fran chopped 6 carrots. ___
5. Bill chopped 8 apples. + ___

[ ] ________________

B

How many people run fast?

1. 3 people run fast. ___
2. 2 people run fast. ___
3. 5 people run fast. ___
4. 9 people run fast. ___
5. 6 people drive fast. + ___

[ ] ________________

C

How many cans did the man carry?

1. The man dropped 4 cans. ___
2. The man carried 2 cans. ___
3. The man carried 2 cans. ___
4. The man carried 3 boxes. + ___

[ ] ________________

Part 5 continues on the next page.

D

How many castles did the girl build in the sand?

1. The girl built 5 castles. ___
2. The girl built 5 shelves. ___
3. The girl built 2 castles. ___
4. The girl found 3 castles. ___
5. The boy built 1 castle. + ___

[ ] ________________

6

A 6 + 452 + 30 = ______

__ __ __
__ __ __
+__ __ __

B 45 + 203 + 21 = ______

__ __ __
__ __ __
+__ __ __

7

A
```
  325
   30
    4
+ 140
```

B
```
  32
  51
  20
+ 42
```

C
```
  53
   1
  40
+  5
```

D
```
  52
  43
  10
+ 81
```

E
```
    8
  310
   20
+ 401
```

Lesson 25

Test	+	Facts	+	Problems	+	Bonus	=	TOTAL

1

2 + 10	1 + 10	5 + 10	8 + 10	3 + 10	9 + 10	6 + 10	1 + 10
7 + 10	4 + 10	6 + 10	2 + 10	5 + 10	8 + 10	3 + 10	2 + 10

2

10 + 7	4 + 5	10 + 4	5 + 3	10 + 7	2 + 4	10 + 9	10 + 5
10 + 3	2 + 5	1 + 7	10 + 2	10 + 8	10 + 1	10 + 6	4 + 5

3

A	B	C	D	E	F
18	53	36	14	17	24

4

A	B	C	D	E
3456	4546	2538	1324	8260

5

A	B	C	D	E	F
42 + 5	73 + 2	45 + 3	21 + 4	11 + 6	24 + 3

6

5 + 9	5 + 6	5 + 9	5 + 8	5 + 7	5 + 9	5 + 8	5 + 7
5 + 5	5 + 9	5 + 6	5 + 8	5 + 5	5 + 7	5 + 9	5 + 8

7

A

How many flowers did Don water?

1. Don cut 16 flowers. ___
2. Don watered 13 flowers. ___
3. Don watered 15 flowers. ___
4. Don watered 10 dogs. ___
5. Chita watered 9 flowers. + ___

[ ] ________________

B

How many pins did Lisa buy?

1. Lisa found 28 pins. ___
2. Lisa bought 24 pins. ___
3. Pam bought 14 pins. ___
4. Lisa bought 11 pins. ___
5. Lisa bought 21 pins. + ___

[ ] ________________

C

Find out how many hats Fred wore.

1. Jay wore 8 hats. ___
2. Lori wore 2 hats. ___
3. Fred wore 5 hats. ___
4. Fred wore 3 socks. ___
5. Fred wore 4 hats. + ___

[ ] ________________

Part 7 continues on the next page.

D

How many students watched TV?

1. 8 teachers watched TV. ___
2. 10 parents watched TV. ___
3. 4 students watched TV. ___
4. 4 students watched the dogs. ___
5. 2 students watched TV. + ___

[ ] ____________

8

A $40 + 3 + 115 =$ ______

B $8 + 301 + 20 =$ ______

9

A
$$\begin{array}{r} 202 \\ 3 \\ +\ 411 \\ \hline \end{array}$$

B
$$\begin{array}{r} 3 \\ 5 \\ 10 \\ +\ 121 \\ \hline \end{array}$$

C
$$\begin{array}{r} 2 \\ 53 \\ 54 \\ +\ 30 \\ \hline \end{array}$$

D
$$\begin{array}{r} 52 \\ 45 \\ 10 \\ +\ 61 \\ \hline \end{array}$$

E
$$\begin{array}{r} 352 \\ 100 \\ 200 \\ +\ 143 \\ \hline \end{array}$$

Lesson 26

Facts + Problems + Bonus = TOTAL

1

5 + 7	5 + 9	5 + 8	5 + 6	5 + 9	5 + 7	5 + 9	5 + 6
5 + 6	5 + 7	5 + 9	5 + 6	5 + 8	5 + 5	5 + 7	5 + 9

2

A 12 B 47 C 15 D 21 E 18 F 24

3

A 4285 B 3624 C 8153 D 9284 E 7295

4

A 35 + 2 B 52 + 2 C 41 + 5 D 75 + 4 E 11 + 3 F 25 + 3

5

7 + 3	7 + 1	7 + 2	7 + 1	7 + 3	7 + 2	7 + 1	7 + 3

6

4 + 10	8 + 10	5 + 10	4 + 5	3 + 10	2 + 5	1 + 8	7 + 10
10 + 3	3 + 5	1 + 7	5 + 4	2 + 4	5 + 3	2 + 3	0 + 5

7

A

Tom climbed 35 steps. ___

Tom climbed 22 steps. ___

Tom climbed 16 trees. ___

Jim climbed 21 steps. + ___

Add the number of steps Tom climbed. [ ] ________________

B

Lucia painted 20 plates on Monday. ___

She painted 15 bowls on Tuesday. ___

She painted 7 cups on Wednesday. ___

She painted 14 glasses on Thursday. ___

Friday she painted 25 plates. + ___

Add the number of plates that Lucia painted. [ ] ________________

C

The man sent 31 letters. ___

The man sent 22 letters. ___

He wrote 26 letters. ___

He sent 11 letters. ___

The woman sent 15 letters. + ___

How many letters did the man send? [ ] ________________

Part 7 continues on the next page.

D

The man used 53 brushes. ___ ___

The man used 21 brushes. ___ ___

The man cleaned 30 brushes. ___ ___

The man used 22 pencils. ___ ___

The man used 11 brushes. + ___ ___

How many brushes did the man use? [ ] ------------------

E

The truck passed 20 runners. ___ ___

The car passed 25 runners. ___ ___

Then the car passed 6 trucks. ___ ___

The car passed 15 runners. + ___ ___

How many runners did the car pass? [ ] ------------------

8

A 300 + 20 + 31 = ______

___ ___ ___

___ ___ ___

+ ___ ___ ___

B 5 + 314 + 20 = ______

___ ___ ___

___ ___ ___

+ ___ ___ ___

C 6 + 452 + 30 = ______

___ ___ ___

___ ___ ___

+ ___ ___ ___

D 45 + 203 + 21 = ______

___ ___ ___

___ ___ ___

+ ___ ___ ___

9

A
```
 340
   5
 250
+  3
────
```

B
```
 80
 10
 12
+75
───
```

C
```
 50
 50
 50
+10
───
```

D
```
   1
 244
  10
+ 31
────
```

E
```
 43
 54
 10
+41
───
```

F
```
 410
 534
   1
+ 23
────
```

G
```
 10
 40
 50
+60
───
```

H
```
 503
   2
  20
+ 51
────
```

I
```
 52
 51
 40
+11
───
```

J
```
 30
 50
 10
+10
───
```

Lesson 27

Facts	+	Problems	+	Bonus	=	TOTAL

1

5 + 9	5 + 7	5 + 9	5 + 8	5 + 6	5 + 8	5 + 10	5 + 7
5 + 9	5 + 6	5 + 8	5 + 7	5 + 10	5 + 8	5 + 7	5 + 9

2

8 + 10	3 + 10	10 + 5	4 + 2	7 + 10	1 + 10	5 + 3	6 + 10
2 + 10	5 + 2	4 + 10	9 + 10	0 + 9	5 + 10	3 + 2	10 + 6

3

A 14 **B** 32 **C** 18 **D** 26 **E** 19 **F** 21

4

A 3279 **B** 146 **C** 6571 **D** 432 **E** 2384

5

75 + 4	11 + 3	45 + 2	20 + 5	12 + 5	32 + 2	42 + 5	25 + 4

6

7 + 4	7 + 1	7 + 4	7 + 2	7 + 4	7 + 3	7 + 2	7 + 4

7

A

The woman climbed 12 hills. ____

The woman climbed 15 mountains. ____

The woman flew over 7 mountains. ____

The woman crossed 26 rivers. ____

The woman climbed 16 trees. ____

The woman climbed 10 mountains. + ____

Add the mountains the woman climbed. [ ] ________________

B

Luis cooked breakfast 10 times. ____

Joe cooked breakfast 14 times. ____

Sharon cooked breakfast 13 times. ____

Luis made candy 15 times. ____

Luis cooked breakfast 11 times. ____

Luis cooked breakfast 31 times. + ____

Find out how many times Luis cooked breakfast. [ ] ________________

C

30 bugs crawl. ____

40 snakes crawl. ____

20 bugs crawl. ____

50 bugs crawl. ____

10 bugs hop. ____

40 bugs crawl. + ____

How many bugs crawl? [ ] ________________

Part 7 continues on the next page.

D

The red team won 8 games. ___

The blue team won 4 games. ___

The blue team tied 1 game. ___

The blue team won 1 trophy. ___

The blue team won 5 games. + ___

How many games did the blue team win? [ ] ________________

8

A 11 + 104 + 23 = ______

___ ___ ___
___ ___ ___
+ ___ ___ ___

B 5 + 360 + 14 = ______

___ ___ ___
___ ___ ___
+ ___ ___ ___

9

A
$$\begin{array}{r} 542 \\ 13 \\ 24 \\ +\ 10 \\ \hline \end{array}$$

B
$$\begin{array}{r} 40 \\ 223 \\ 103 \\ +\ 2 \\ \hline \end{array}$$

C
$$\begin{array}{r} 50 \\ 40 \\ 10 \\ +60 \\ \hline \end{array}$$

D
$$\begin{array}{r} 47 \\ 200 \\ 11 \\ +\ 20 \\ \hline \end{array}$$

E
$$\begin{array}{r} 51 \\ 52 \\ 40 \\ +12 \\ \hline \end{array}$$

Lesson 28

Facts	+	Problems	+	Bonus	=	TOTAL

1

$5+7$ $5+9$ $5+6$ $5+10$ $5+8$ $5+7$ $5+6$ $5+9$

$5+8$ $5+10$ $5+6$ $5+9$ $5+7$ $5+10$ $5+9$ $5+6$

2

A 15   B 36   C 12   D 17   E 20   F 14

3

$4+10$ $3+5$ $2+10$ $10+7$ $5+4$ $3+10$ $2+5$ $1+4$

$2+5$ $0+7$ $5+2$ $3+5$ $2+3$ $5+10$ $6+0$ $5+3$

4

A 3856   B 4258   C 326   D 1420

E 350   F 4263   G 9412   H 412

5

A $43+2$ B $31+5$ C $73+2$ D $62+5$ E $52+2$ F $11+3$ G $41+7$

6

7	7	7	7	7	7	7	7	7
+ 4	+ 2	+ 5	+ 4	+ 3	+ 5	+ 3	+ 2	+ 4

7

A	B	C	D
☐	☐	☐	☐
39	19	23	29
19	25	22	59
22	26	35	69
+ 15 (25)	+ 13 (23)	+ 14 (14)	+ 15 (32)

8

A

1. Ed bought 2 apples. ___
2. Ed found 3 apples. ___
3. Drew found 4 apples. ___
4. Ed bought 2 oranges. ___
5. Ed saw 5 apples. ___
6. Ed bought 3 candy bars. + ___

Add the apples Ed got. ☐ ____________________

B

1. Judy walked to school 3 times. ___
2. Judy rode her bike to school 1 time. ___
3. Judy ran to the store 2 times. ___
4. Judy took the bus to school 5 times. ___
5. Andy walked to school 4 times. + ___

Add the times Judy went to school. ☐ ____________________

Part 8 continues on the next page.

C

1. Dave ate hot dogs for 2 hours. ___
2. Dave pulled weeds for 2 hours. ___
3. Carmen cut grass for 1 hour. ___
4. Dave washed cars for 2 hours. ___
5. Dave watched TV for 1 hour. + ___

How many hours did Dave work? [ ] __________________

9

A 2 1 + 2 0 4 + 3 = ______

__ __ __
__ __ __
\+ __ __ __

B 6 + 3 5 1 + 3 1 1 = ______

__ __ __
__ __ __
\+ __ __ __

Lesson 29

Facts	+	Problems	+	Bonus	=	TOTAL

1

7	7	7	7	7	7	7	7
+ 2	+ 5	+ 3	+ 1	+ 4	+ 3	+ 5	+ 2

2

A 4236 **B** 8150 **C** 425 **D** 7111

E 711 **F** 250 **G** 4250

3

A	B	C	D	E
32	11	15	25	11
+ 4	+ 4	+ 3	+ 4	+ 4

4

A	B	C	D
☐	☐	☐	☐
29	43	38	16
19	5	9	33
53	26	39	54
+ 25 (26)	+ 1 (15)	+ 8 (34)	+ 64 (17)

5

A ☐ { 5 ________ / 8 ________

B ☐ { 5 ________ / 6 ________

C ☐ { 5 ________ / 9 ________

D ☐ { 5 ________ / 7 ________

6

5 + 8	3 + 10	5 + 6	5 + 9	4 + 10	2 + 2	5 + 10	5 + 7
3 + 5	14 + 1	5 + 7	5 + 5	10 + 5	5 + 9	3 + 10	5 + 3

7

A

1. Herman found 3 sheep.
2. Herman bought 7 sheep.
3. Herman lost 6 sheep.
4. Herman gave away 9 sheep.
5. Herman bought 4 sheep.

Add the sheep that Herman got.

B

1. Alma put 5 jars in the box.
2. Alma took 3 shoes from the box.
3. Alma dumped 5 letters in the box.
4. Alma placed 6 cans in the box.
5. Alma pulled 8 pencils from the box.

Add the things that made the box heavier.

Part 7 continues on the next page.

C

1. Mr. Clearwater bought 20 CDs.
2. Mr. Clearwater won 20 CDs.
3. Mr. Begay bought 30 CDs.
4. Mr. Clearwater sold 50 CDs.
5. Mr. Clearwater found 10 CDs.
6. Mr. Clearwater sold 40 CDs.

Add the CDs Mr. Clearwater got.

D

1. Jess has 5 cowboy hats.
2. Jess has 5 straw hats.
3. Jess has 1 coin.
4. His sister has 3 hats.
5. Jess has 4 blue hats.

How many hats does Jess have?

8

A 321 + 50 + 4 = ______

B 4 + 352 + 20 = ______

9

A	B	C	D	E
321	720	50	53	340
40	40	90	71	25
205	100	10	10	100
+ 110	+ 110	+ 10	+ 5	+ 514

Lesson 30

Facts	+	Problems	+	Bonus	=	TOTAL

1

7 + 3	7 + 1	7 + 5	7 + 4	7 + 2	7 + 4	7 + 1	7 + 3	7 + 5
7 + 2	7 + 5	7 + 4	7 + 3	7 + 1	7 + 5	7 + 3	7 + 2	7 + 4

2

A { 5 ________ 9 ________

B { 5 ________ 6 ________

C { 5 ________ 8 ________

D { 5 ________ 7 ________

3

A	B	C	D	E
15 + 3	12 + 5	41 + 7	11 + 3	74 + 5

4

5 + 6	5 + 8	5 + 3	10 + 3	5 + 9	5 + 5	5 + 7	4 + 10
4 + 5	8 + 10	5 + 7	2 + 2	0 + 7	6 + 10	5 + 6	5 + 8

5

A 4000

B 400

C 318

D 1150

E 2314

F 107

G 2360

6

A
```
  35
  59
+ 11
```

B
```
  11
  29
  49
+ 11
```

C
```
  45
  25
+ 54
```

D
```
  81
  14
+ 25
```

7

A

1. Sharon lost 4 pencils.
2. Sharon gave away 9 pencils.
3. Sharon found 2 pieces of paper.
4. Sharon threw away 5 crayons.
5. Sharon bought 8 pens.

Add the things that Sharon got.

B

1. Pablo took 9 cups from the pile.
2. Pablo placed 14 pots on the pile.
3. Pablo set 5 plates on the pile.
4. Pablo took 4 glasses from the pile.
5. Pablo took 10 cups from the pile.

Add the things that made the pile bigger.

Part 7 continues on the next page.

C

1. Team 1 won 13 games in May.
2. Team 2 won 14 games in May.
3. Team 1 won 20 games in June.
4. Team 1 lost 8 games in June.
5. Team 1 won 21 games in July.

Add the games team 1 won.

D

1. The boys carried boxes 52 times.
2. The boys raked leaves 42 times.
3. The boys chopped wood 11 times.
4. The boys played catch 18 times.
5. The girls raked leaves 24 times.
6. The boys piled wood 12 times.

How many times did the boys do work?

Lesson 31

Test + Facts + Problems + Bonus = TOTAL

1

7	7	7	7	7	7	7	7
+ 5	+ 4	+ 3	+ 5	+ 2	+ 4	+ 1	+ 5

7	7	7	7	7	7	7	7
+ 3	+ 5	+ 1	+ 4	+ 2	+ 4	+ 3	+ 5

2

5	7	9	8	6	9	8	7
+ 5	+ 5	+ 5	+ 5	+ 5	+ 5	+ 5	+ 5

8	10	7	6	9	7	10	8
+ 5	+ 5	+ 5	+ 5	+ 5	+ 5	+ 5	+ 5

3

A 508 **B** 3740 **C** 2851 **D** 1418

E 231 **F** 3266 **G** 4333

4

A 74 + 5 = ______ **B** 15 + 3 = ______ **C** 12 + 3 = ______

D 11 + 4 = ______ **E** 35 + 2 = ______ **F** 41 + 7 = ______

G 11 + 5 = ______ **H** 14 + 2 = ______

5

5 + 6	1 + 2	3 + 10	5 + 4	5 + 8	1 + 8	3 + 2
0 + 6	5 + 5	10 + 8	2 + 4	4 + 10	4 + 2	0 + 3
5 + 7	10 + 2	5 + 9	5 + 3	2 + 3	2 + 5	4 + 5

6

A

1. Miss Kickingbird put 2 cans of paint in her car.
2. Miss Kickingbird took 4 bags of food out of her car.
3. Miss Kickingbird used 5 rags to clean her car.
4. Miss Kickingbird put 2 boxes in her car.
5. 5 children got out of Miss Kickingbird's car.

Add the things that made Miss Kickingbird's car heavier.

B

1. Karl bought 40 hamsters.
2. He was given 20 hamsters.
3. He found 20 snakes.
4. He found 10 hamsters.
5. He lost 10 hamsters.

Add the hamsters that Karl got.

Part 6 continues on the next page.

C

1. Shelly was given 2 glasses.
2. She broke 4 glasses.
3. She found 2 glasses.
4. She bought 5 glasses.
5. She was given 1 glass.
6. She bought 5 plates.

Add the glasses that Shelly got.

D

1. Justin cleaned 32 shirts.
2. He cleaned 13 dresses.
3. He cleaned 22 skirts.
4. He sold 19 socks.
5. He cleaned 16 dishes.

Add the pieces of clothing that Justin cleaned.

7

A
$$\begin{array}{r} 31 \\ 19 \\ +\,22 \\ \hline \end{array}$$

B
$$\begin{array}{r} 21 \\ 24 \\ 55 \\ +\,43 \\ \hline \end{array}$$

C
$$\begin{array}{r} 75 \\ 25 \\ 15 \\ +\,12 \\ \hline \end{array}$$

D
$$\begin{array}{r} 33 \\ 122 \\ +\,415 \\ \hline \end{array}$$

E
$$\begin{array}{r} 51 \\ 19 \\ 24 \\ +\,11 \\ \hline \end{array}$$

8

A $4 + 31 + 210 =$ ______

B $62 + 5 + 310 =$ ______

Lesson 32

Facts	+	Problems	+	Bonus	=	TOTAL

1

8 + 5	6 + 5	7 + 5	9 + 5	8 + 5	6 + 5	7 + 5	9 + 5
5 + 5	9 + 5	6 + 5	8 + 5	6 + 5	9 + 5	7 + 5	8 + 5

2

A [ ] { 7 ____ / 2 ____ }

B [ ] { 7 ____ / 5 ____ }

C [ ] { 7 ____ / 3 ____ }

D [ ] { 7 ____ / 1 ____ }

E [ ] { 7 ____ / 4 ____ }

F [ ] { 7 ____ / 3 ____ }

3

7 + 4	4 + 5	7 + 5	5 + 9	7 + 3	5 + 8	5 + 4	10 + 4
0 + 7	3 + 5	7 + 2	5 + 7	2 + 10	5 + 10	0 + 7	5 + 9

4

A 7214

B 9382

C 451

D 4160

E 380

F 5246

G 1328

5

A $12 + 4 =$ ______

B $13 + 5 =$ ______

C $11 + 6 =$ ______

D $12 + 3 =$ ______

E $17 + 2 =$ ______

F $12 + 5 =$ ______

G $42 + 5 =$ ______

H $32 + 4 =$ ______

I $13 + 4 =$ ______

J $53 + 4 =$ ______

K $11 + 4 =$ ______

L $12 + 4 =$ ______

6

A

Susan bought 4 cakes. She gave away 2 cakes. Her friend gave her 5 cakes. She found 1 cake. She bought 2 pies. Add the cakes that Susan got.

B

Julia played cards for 2 hours. She played ball for 3 hours. She did homework for 2 hours. Julia jumped rope for 2 hours. Add the hours that Julia played.

Part 6 continues on the next page.

C

Hiro is a nurse. He takes care of patients 20 times. He makes beds 40 times. He takes care of patients 30 times. He answers the phone 10 times. He takes care of patients 50 times. Add the times Hiro takes care of patients.

D

Rosa bought 25 rocks for her collection. Trudy bought 14 rocks. Rosa found 30 rocks. She gave away 10 rocks. She found 8 bricks. How many rocks did Rosa get?

7

A
$$\begin{array}{r} 271 \\ 109 \\ 305 \\ +\ 11 \\ \hline \end{array}$$

B
$$\begin{array}{r} 127 \\ 353 \\ 4 \\ +\ 11 \\ \hline \end{array}$$

C
$$\begin{array}{r} 129 \\ 21 \\ 219 \\ +111 \\ \hline \end{array}$$

D
$$\begin{array}{r} 25 \\ 54 \\ 11 \\ +18 \\ \hline \end{array}$$

E
$$\begin{array}{r} 91 \\ 54 \\ 15 \\ +\ 3 \\ \hline \end{array}$$

8

A $42 + 317 + 20 =$ ______

B $4 + 2 + 51 =$ ______

Lesson 33

Facts	+	Problems	+	Bonus	=	TOTAL

1

10	6	8	7	9	6	8	7
+ 5	+ 5	+ 5	+ 5	+ 5	+ 5	+ 5	+ 5

8	6	7	8	10	7	9	8
+ 5	+ 5	+ 5	+ 5	+ 5	+ 5	+ 5	+ 5

2

A ☐ { 7 ______ ; 3 ______

B ☐ { 7 ______ ; 2 ______

C ☐ { 7 ______ ; 5 ______

D ☐ { 7 ______ ; 4 ______

E ☐ { 7 ______ ; 1 ______

F ☐ { 7 ______ ; 3 ______

3

5	7	5	5	7	5	3	5
+ 7	+ 4	+ 9	+ 10	+ 3	+ 6	+ 10	+ 4

5	8	7	5	7	3	4	10
+ 8	+ 10	+ 4	+ 9	+ 5	+ 5	+ 10	+ 5

4

A 12 + 5 = ______

B 11 + 4 = ______

C 43 + 5 = ______

D 72 + 4 = ______

E 15 + 3 = ______

F 11 + 3 = ______

G 13 + 2 = ______

H 94 + 3 = ______

I 76 + 2 = ______

5

A 4056

B 5207

C 8024

D 2902

E 4908

F 7067

6

A
$$\begin{array}{r} 25 \\ +\ \ 9 \\ \hline 4 \end{array}$$

B
$$\begin{array}{r} 65 \\ +\ \ 9 \\ \hline 4 \end{array}$$

C
$$\begin{array}{r} 85 \\ +\ \ 9 \\ \hline 4 \end{array}$$

D
$$\begin{array}{r} 15 \\ +\ \ 9 \\ \hline 4 \end{array}$$

7

A

Leroy was given 4 eggs. He scrambled 1 egg for lunch. He bought 6 eggs. Someone gave him 1 apple. He bought 6 eggs. Add the eggs Leroy got.

B

The tractor dug rows for 21 hours. It cut grass for 32 hours. It loaded corn for 22 hours. It sat in the field for 41 hours. It planted seeds for 13 hours. How many hours did the tractor work?

Part 7 continues on the next page.

C

Maya fed 5 horses. She rode 3 horses. She fed 5 pigs. She fed 6 cats. She ran after 8 chickens. Add the animals that Maya fed.

D

Nancy made 43 shirts. She washed 25 shirts. She made 12 hats. She made 53 jackets. She made 3 sandwiches. Add the pieces of clothing that Nancy made.

8

A
$$\begin{array}{r} 9 \\ 31 \\ 5 \\ +\ 211 \\ \hline \end{array}$$

B
$$\begin{array}{r} 86 \\ 11 \\ 11 \\ +\ 70 \\ \hline \end{array}$$

C
$$\begin{array}{r} 451 \\ 9 \\ 209 \\ +\ \ 11 \\ \hline \end{array}$$

D
$$\begin{array}{r} 724 \\ 21 \\ 234 \\ +\ \ 11 \\ \hline \end{array}$$

E
$$\begin{array}{r} 55 \\ 50 \\ 23 \\ +\ 10 \\ \hline \end{array}$$

9

A $25 + 302 + 51 =$ ______

B $7 + 182 + 210 =$ ______

Lesson 34

Facts	+	Problems	+	Bonus	=	TOTAL

1

$9 + 5$	$6 + 5$	$8 + 5$	$10 + 5$	$7 + 5$	$6 + 5$	$9 + 5$	$7 + 5$
$7 + 5$	$10 + 5$	$8 + 5$	$9 + 5$	$6 + 5$	$7 + 5$	$10 + 5$	$8 + 5$

2

A	B	C	D	E	F
$47 + 5$	$67 + 5$	$37 + 5$	$17 + 5$	$67 + 5$	$27 + 5$

3

$5 + 7$	$2 + 7$	$4 + 7$	$3 + 7$	$1 + 7$	$2 + 7$	$4 + 7$	$1 + 7$
$4 + 7$	$2 + 7$	$5 + 7$	$4 + 7$	$1 + 7$	$3 + 7$	$4 + 7$	$2 + 7$

4

$5 + 8$	$5 + 6$	$7 + 4$	$5 + 10$	$7 + 3$	$5 + 7$	$5 + 9$	$8 + 10$
$5 + 2$	$10 + 3$	$5 + 3$	$3 + 10$	$7 + 2$	$5 + 10$	$7 + 5$	$10 + 7$

5

A 11 + 3 = ______ B 12 + 5 = ______ C 77 + 2 = ______

D 14 + 5 = ______ E 41 + 5 = ______ F 13 + 5 = ______

G 52 + 2 = ______ H 32 + 5 = ______ I 11 + 5 = ______

J 12 + 3 = ______ K 25 + 3 = ______ L 43 + 5 = ______

6

A 4006 B 5027 C 7004 D 2902

E 4098 F 7007

A

Luz walked to the library 2 times. She ran to the library 2 times. She rode her bike to the library 5 times. She crawled into bed 3 times. She skipped to the library 6 times. She walked to the library 1 more time. How many times did Luz go to the library?

B

The woman sold 11 dresses. She sold 13 hats. She sold 31 coats. She fixed 12 shoes. She sold 22 pencils. She sold 23 shirts. Add the pieces of clothing the woman sold.

Part 7 continues on the next page.

C

The men and women moved bricks for 12 hours. They dug for 31 hours. They shoveled sand for 15 hours. They rested for 5 hours. They cut wood for 51 hours. They talked to each other for 21 hours. Add the hours the men and women worked.

8

A
$$\begin{array}{r} 345 \\ 107 \\ 11 \\ +\ 200 \\ \hline \end{array}$$

B
$$\begin{array}{r} 41 \\ 79 \\ 19 \\ +\ 11 \\ \hline \end{array}$$

C
$$\begin{array}{r} 75 \\ 14 \\ 10 \\ +\ 10 \\ \hline \end{array}$$

D
$$\begin{array}{r} 125 \\ 200 \\ 105 \\ +\ 558 \\ \hline \end{array}$$

E
$$\begin{array}{r} 45 \\ 13 \\ 70 \\ +\ 10 \\ \hline \end{array}$$

9

A $47 + 3 + 120 =$ ______

B $4 + 50 + 130 =$ ______

Lesson 35

Test	+	Facts	+	Problems	+	Bonus	=	TOTAL

1

$4 + 7$ $2 + 7$ $5 + 7$ $1 + 7$ $3 + 7$ $5 + 7$ $4 + 7$ $3 + 7$

$5 + 7$ $3 + 7$ $5 + 7$ $4 + 7$ $2 + 7$ $5 + 7$ $3 + 7$ $4 + 7$

2

A $65 + 8$ **B** $35 + 8$ **C** $85 + 8$ **D** $15 + 8$ **E** $65 + 8$ **F** $25 + 8$

3

$7 + 5$ $3 + 10$ $8 + 5$ $7 + 3$ $9 + 5$ $7 + 4$ $7 + 5$ $6 + 5$

$6 + 10$ $5 + 9$ $7 + 4$ $7 + 10$ $5 + 8$ $5 + 7$ $5 + 6$ $10 + 5$

4

A $12 + 3 =$ ______   **B** $65 + 4 =$ ______   **C** $11 + 7 =$ ______

D $12 + 2 =$ ______   **E** $15 + 3 =$ ______   **F** $21 + 8 =$ ______

G $45 + 3 =$ ______   **H** $24 + 2 =$ ______   **I** $12 + 4 =$ ______

J $34 + 5 =$ ______   **K** $11 + 4 =$ ______   **L** $13 + 2 =$ ______

5

A 5009

B 9035

C 8024

D 7406

E 9004

F 8201

G 5034

6

8	6	8	9	6	9	8	7
+ 2	+ 2	+ 2	+ 2	+ 2	+ 2	+ 2	+ 2

7

A	B	C	D	E
345	241	49	234	44
207	104	51	100	41
141	48	59	205	17
+ 100	+ 200	+ 11	+ 150	+ 71

8

A

Terry loaded 5 pieces of furniture into his car. He put 8 plants in his car. He took 6 suits out of his car. 9 boys got out of his car. He took 5 clothing bags out of his car. How many things made his car heavier?

B

20 girls were in the bus. 10 men were in the bus. 20 boys were in the car. 10 dogs were in the bus. 20 women were in the bus. How many people were in the bus?

Part 8 continues on the next page.

C

Jerry saw 4 robins. Omar saw 8 robins. Jerry saw 1 robin. Jerry saw 7 woodpeckers. How many robins did Jerry see?

D

Ram bought 20 plants. He watered 30 plants. He saw 40 plants. Ram sold 50 plants. He found 10 plants. Add the plants that Ram got.

E

90 people walked to the campfire. 40 people left the campfire. 10 people walked home from the campfire. 20 people left the campfire. 10 people drove to the campfire. How many people came to the campfire?

9

A 7 + 315 + 4 = ______

B 6 + 320 + 30 = ______

Lesson 36

Facts	+	Problems	+	Bonus	=	TOTAL

1

A	B	C	D	E	F
5	8	4	7	2	6
+ 9	+ 9	+ 9	+ 9	+ 9	+ 9

2

A 4064 B 5002 C 9400 D 8006

E 6010 F 8000 G 5407

3

A	B	C	D	E	F
75	15	57	37	15	65
+ 9	+ 9	+ 5	+ 5	+ 8	+ 8

4

3	1	4	2	5	3	5	4
+ 7	+ 7	+ 7	+ 7	+ 7	+ 7	+ 7	+ 7
5	4	3	5	2	4	1	3
+ 7	+ 7	+ 7	+ 7	+ 7	+ 7	+ 7	+ 7

5

8	6	7	9	6	7	9	8
+ 2	+ 2	+ 2	+ 2	+ 2	+ 2	+ 2	+ 2

6

7 + 10	7 + 5	7 + 4	9 + 5	3 + 10	8 + 5	10 + 5	6 + 5
6 + 10	5 + 8	10 + 4	5 + 6	9 + 2	5 + 7	6 + 2	5 + 3

7

A 17 + 2 = ______ **B** 61 + 4 = ______ **C** 13 + 4 = ______

D 12 + 5 = ______ **E** 41 + 7 = ______ **F** 15 + 3 = ______

G 43 + 5 = ______ **H** 34 + 2 = ______ **I** 73 + 2 = ______

J 12 + 4 = ______ **K** 11 + 5 = ______ **L** 23 + 5 = ______

8

A

37 cows were in the barn. 2 horses were in the barn. 14 pigs were in the field. 20 cats were in the barn. 2 tractors were in the barn. How many animals were in the barn?

B

Willie ran 21 kilometers. Tammy ran 45 kilometers. Holly ran 25 kilometers. Jane ran 6 kilometers. Inéz rode her bike 14 kilometers. How many kilometers did the girls run?

Part 8 continues on the next page.

C

The blue team won 27 games in April. The red team won 14 games in April. The blue team lost 9 games in April. The blue team won 5 games in May. The blue team won 10 games in June. How many games did the blue team win?

D

Amy put 13 rag dolls into the closet. She took 4 stuffed monkeys out of the closet. She pulled 25 tin soldiers out of the closet. Amy shoved 4 fans into the closet. She pulled 2 floppy hats out of the closet. How many things did Amy take from the closet?

9

A
$$\begin{array}{r} 351 \\ 154 \\ 242 \\ +\ 111 \\ \hline \end{array}$$

B
$$\begin{array}{r} 213 \\ 192 \\ 393 \\ +\ 111 \\ \hline \end{array}$$

C
$$\begin{array}{r} 971 \\ 453 \\ 232 \\ +\ \ 41 \\ \hline \end{array}$$

D
$$\begin{array}{r} 394 \\ 111 \\ 323 \\ +\ 111 \\ \hline \end{array}$$

E
$$\begin{array}{r} 671 \\ 132 \\ 21 \\ +\ 135 \\ \hline \end{array}$$

F
$$\begin{array}{r} 471 \\ 543 \\ 222 \\ +\ 211 \\ \hline \end{array}$$

10

A $5 + 204 + 30 =$ ______

B $23 + 5 + 211 =$ ______

Lesson 37

Facts	+	Problems	+	Bonus	=	TOTAL

1

A	B	C	D	E	F
$6 + 9$	$8 + 9$	$4 + 9$	$5 + 9$	$7 + 9$	$3 + 9$

2

A 4006

B 508

C 300

D 3000

E 502

F 5020

G 5200

3

A	B	C	D	E	F
$15 + 8$	$65 + 8$	$37 + 5$	$17 + 5$	$75 + 9$	$15 + 9$

4

A

$5 + 7$, $2 + 7$, $3 + 7$, $4 + 7$, $2 + 7$, $5 + 7$, $3 + 7$

B

$10 + 7$, $4 + 7$, $2 + 7$, $5 + 7$, $3 + 7$, $2 + 7$, $4 + 7$

5

$8 + 5$, $7 + 4$, $7 + 5$, $6 + 5$, $9 + 5$, $10 + 5$, $4 + 5$, $5 + 7$

$9 + 5$, $7 + 3$, $5 + 8$, $3 + 5$, $8 + 5$, $6 + 5$, $5 + 3$, $9 + 5$

6

A 12 + 4 = ______

B 51 + 7 = ______

C 72 + 3 = ______

D 13 + 5 = ______

E 11 + 4 = ______

F 12 + 5 = ______

G 43 + 5 = ______

H 14 + 5 = ______

I 61 + 8 = ______

7

A
$$\begin{array}{r} 395 \\ 145 \\ 212 \\ +\ 101 \\ \hline \end{array}$$

B
$$\begin{array}{r} 297 \\ 95 \\ 310 \\ +\ 203 \\ \hline \end{array}$$

C
$$\begin{array}{r} 227 \\ 252 \\ 310 \\ +\ 120 \\ \hline \end{array}$$

D
$$\begin{array}{r} 981 \\ 553 \\ 421 \\ +\ 142 \\ \hline \end{array}$$

E
$$\begin{array}{r} 205 \\ 195 \\ +\ 111 \\ \hline \end{array}$$

F
$$\begin{array}{r} 65 \\ 58 \\ 41 \\ +\ 10 \\ \hline \end{array}$$

8

A

Blanca ran for 35 minutes in the morning. She jumped rope for 20 minutes in the afternoon. She ran for 120 minutes in the evening. The next morning she did 150 knee bends. That night she ran for 40 minutes. How many minutes did Blanca run?

B

In the food store there were 114 oranges and 72 apples on a shelf. There were 92 boxes of sugar on a shelf. There were 3 watermelons on the floor. There were 10 bananas on a shelf. Find out how many pieces of fruit were in the store.

Part 8 continues on the next page.

C

Sol saw 38 green trucks. He saw 27 red cars. He saw 15 blue cars. He saw 21 white cars. Sol's friend saw 30 blue cars. How many cars did Sol see?

D

Leona took 5 puppies for a walk. She took 3 of her sisters for a walk. She took 10 dogs for a walk. Jim took 4 cats for a walk. How many pets did Leona take for a walk?

E

Mr. Arrowhead put 10 tires in the garage. He put 40 bags of grass seed in the garage. He took 20 cans of paint out of the garage. He carried 20 kittens out of the garage. He chased 10 dogs out of the garage. How many things were added to Mr. Arrowhead's garage?

A 3 + 67 + 54 = ______

B 8 + 200 + 40 = ______

Lesson 38

Facts + Problems + Bonus = TOTAL

1

A $6 + 9$

B $4 + 9$

C $9 + 9$

D $2 + 9$

E $8 + 9$

F $5 + 9$

2

A 2045

B 2405

C 8001

D 920

E 4054

F 1099

G 5002

3

A $87 + 5$

B $17 + 5$

C $65 + 8$

D $15 + 8$

E $39 + 5$

F $19 + 5$

4

$6 + 2$ | $8 + 2$ | $7 + 2$ | $5 + 2$ | $9 + 2$ | $7 + 2$ | $8 + 2$ | $5 + 2$

$9 + 2$ | $7 + 2$ | $10 + 2$ | $6 + 2$ | $8 + 2$ | $5 + 2$ | $9 + 2$ | $6 + 2$

5

$3 + 7$ | $5 + 8$ | $5 + 7$ | $8 + 5$ | $4 + 7$ | $5 + 6$ | $2 + 7$ | $9 + 5$

$5 + 7$ | $3 + 5$ | $5 + 8$ | $7 + 5$ | $7 + 4$ | $5 + 9$ | $7 + 3$ | $5 + 7$

6

A 14 + 5 = ______ **B** 12 + 4 = ______ **C** 62 + 7 = ______

Part 6 continues on the next page.

D $11 + 6 =$ ______ E $11 + 3 =$ ______ F $73 + 5 =$ ______

G $21 + 4 =$ ______ H $34 + 5 =$ ______ I $13 + 2 =$ ______

J $42 + 5 =$ ______ K $12 + 2 =$ ______ L $53 + 4 =$ ______

A

135 people were playing ball. 30 people were building houses. 210 people were fixing cars. 420 people were cleaning the street. 150 people were watching a show. How many people were working?

B

At the zoo, Flo saw 137 monkeys, 25 lions, 13 clowns, and 20 tigers. How many animals did Flo see?

C

Alberto found 5 red birdhouses. He bought 25 green birdhouses. He gave 4 birdhouses to his sister. His friend gave him 103 birdhouses. Add the birdhouses Alberto got.

D

Polly threw away 6 broken bats. She sold 27 green racers. She gave away 35 red racers. She threw away 28 gray racers. How many racers did Polly get rid of?

8

A
$$\begin{array}{r} 78 \\ 325 \\ 12 \\ +\,251 \\ \hline \end{array}$$

B
$$\begin{array}{r} 97 \\ 45 \\ 211 \\ +\,112 \\ \hline \end{array}$$

C
$$\begin{array}{r} 304 \\ 281 \\ 152 \\ +\quad 11 \\ \hline \end{array}$$

D
$$\begin{array}{r} 267 \\ 53 \\ 149 \\ +\,221 \\ \hline \end{array}$$

E
$$\begin{array}{r} 62 \\ 14 \\ 51 \\ +\,113 \\ \hline \end{array}$$

Lesson 39

Facts	+	Problems	+	Bonus	=	TOTAL

1

A $8 + 9$ = ____

B $5 + 9$ = ____

C $7 + 9$ = ____

D $4 + 9$ = ____

E $6 + 9$ = ____

F $3 + 9$ = ____

2

A 600

B 5000

C 2006

D 260

E 5040

F 7100

G 1101

3

A ______

B ______

C ______

D ______

E ______

F ______

4

A [ ] { 7 ________________ ; 2 ________________ }

B [ ] { 10 ________________ ; 2 ________________ }

C [ ] { 8 ________________ ; 2 ________________ }

D [ ] { 5 ________________ ; 2 ________________ }

E [ ] { 6 ________________ ; 2 ________________ }

F [ ] { 9 ________________ ; 2 ________________ }

5

3	5	4	7	5	5	8	5
+ 7	+ 9	+ 7	+ 3	+ 6	+ 8	+ 5	+ 7

7	8	5	6	7	9	7	3
+ 5	+ 8	+ 4	+ 5	+ 4	+ 5	+ 5	+ 7

6

A 17 + 2 = ______ **B** 62 + 5 = ______ **C** 13 + 2 = ______

D 41 + 5 = ______ **E** 14 + 5 = ______ **F** 12 + 3 = ______

G 41 + 6 = ______ **H** 34 + 5 = ______ **I** 37 + 2 = ______

J 12 + 5 = ______ **K** 82 + 7 = ______ **L** 11 + 5 = ______

7

A

Toshio went fishing 3 times. His sister went fishing 2 times. Toshio went fishing 2 more times. He caught 4 fish. How many times did Toshio go fishing?

B

The coach gave away 8 balls. The coach threw away 16 balls. She bought 9 balls. She gave away 10 balls. How many balls did the coach get rid of?

Part 7 continues on the next page.

C

150 children went ice-skating. 70 mothers and fathers went ice-skating. 30 children built a snowman. 200 children went ice fishing. 100 children studied for a test. How many children played outside?

D

Miss Dodge went for a walk in the woods. She saw 105 wild strawberries, 20 dead trees, 31 wild blueberries, 4 wildflowers, 10 cabins, and 3 snakes. How many living things did Miss Dodge see on her walk?

E

Dan's aunt ran up 8 hills. Dan ran up 6 hills. Dan's aunt walked up 5 hills. Dan's aunt drove up 2 hills. She drove down 1 hill. How many hills did Dan's aunt go up?

8

A
$$\begin{array}{r} 108 \\ 14 \\ 13 \\ +\ 122 \\ \hline \end{array}$$

B
$$\begin{array}{r} 99 \\ 35 \\ 532 \\ +\ 122 \\ \hline \end{array}$$

C
$$\begin{array}{r} 250 \\ 70 \\ 20 \\ +\ 210 \\ \hline \end{array}$$

D
$$\begin{array}{r} 602 \\ 55 \\ 25 \\ +\ 112 \\ \hline \end{array}$$

E
$$\begin{array}{r} 563 \\ 351 \\ 135 \\ +\ 110 \\ \hline \end{array}$$

Lesson 40

Facts + Problems + Bonus = TOTAL

1

A $5 + 9$

B $2 + 9$

C $9 + 9$

D $7 + 9$

E $4 + 9$

F $6 + 9$

2

A $700 + 200$

B $100 + 400$

C $200 + 400$

D $300 + 500$

E $500 + 200$

3

A ______

B ______

C ______

D ______

E ______

F ______

4

A ☐ { 9 ______ / 2 ______ }

B ☐ { 7 ______ / 2 ______ }

C ☐ { 5 ______ / 2 ______ }

D ☐ { 10 ______ / 2 ______ }

E ☐ { 8 ______ / 2 ______ }

F ☐ { 6 ______ / 2 ______ }

5

6	4	8	9	5	9	3	7
+ 4	+ 7	+ 2	+ 5	+ 7	+ 2	+ 7	+ 2

8	5	3	6	8	8	9	9
+ 5	+ 6	+ 7	+ 2	+ 5	+ 2	+ 5	+ 2

6

A 12 + 7 = ______ **B** 21 + 5 = ______ **C** 42 + 3 = ______

D 14 + 5 = ______ **E** 53 + 2 = ______ **F** 15 + 4 = ______

G 42 + 6 = ______ **H** 31 + 2 = ______ **I** 12 + 4 = ______

J 56 + 2 = ______ **K** 11 + 4 = ______ **L** 40 + 3 = ______

7

A

150 teachers flew in airplanes. 12 sparrows flew over a mountain. 20 bakers flew in airplanes. 20 doctors flew in airplanes. 17 dogs flew in airplanes. How many people flew in airplanes?

B

The big fish ate 240 flies. The mice ate 29 flies. The birds ate 34 flies. The little fish ate 28 flies. How many flies did the fish eat?

Part 7 continues on the next page.

C

Lynn lost 4 shirts. She was given 2 tools. She bought 3 potatoes. She returned 8 boxes. She bought 2 hats. She picked 3 flowers. She found 7 pencils. How many things did Lynn get?

D

We saw 21 sharks. We saw 150 birds. We saw 20 boats. We saw 310 jellyfish. We were seen by 5 fish. How many fish did we see?

E

The cook made 150 apple pies, 120 cherry pies, 60 baked chickens, and 50 cream pies. He ate 4 cherry pies. How many pies did the cook make?

8

A

$$\begin{array}{r} 403 \\ 157 \\ 454 \\ +\ 130 \\ \hline \end{array}$$

B

$$\begin{array}{r} 925 \\ 425 \\ 251 \\ +\ 134 \\ \hline \end{array}$$

C

$$\begin{array}{r} 905 \\ 964 \\ 151 \\ +\ 310 \\ \hline \end{array}$$

D

$$\begin{array}{r} 270 \\ 230 \\ 540 \\ +\ 810 \\ \hline \end{array}$$

E

$$\begin{array}{r} 37 \\ 25 \\ 13 \\ +\ 51 \\ \hline \end{array}$$

Lesson 41

Test	+	Facts	+	Problems	+	Bonus	=	TOTAL

1

$2 + 9$	$2 + 6$	$2 + 8$	$2 + 5$	$2 + 9$	$2 + 8$	$2 + 5$	$2 + 9$
$2 + 7$	$2 + 5$	$2 + 9$	$2 + 8$	$2 + 6$	$2 + 9$	$2 + 7$	$2 + 6$

2

A	B	C	D	E	F
$37 + 5$	$19 + 5$	$27 + 4$	$12 + 8$	$65 + 8$	$15 + 7$

3

$5 + 9$	$7 + 9$	$6 + 5$	$8 + 9$	$7 + 4$	$6 + 9$	$3 + 9$	$4 + 5$
$5 + 9$	$4 + 9$	$8 + 5$	$3 + 9$	$7 + 5$	$7 + 9$	$4 + 7$	$4 + 9$

4

$6 + 5$	$7 + 2$	$5 + 5$	$6 + 1$	$4 + 5$	$9 + 5$	$1 + 9$	$5 + 6$
$2 + 4$	$7 + 3$	$2 + 7$	$4 + 10$	$5 + 9$	$7 + 4$	$10 + 6$	$5 + 3$

Part 4 continues on the next page.

8	0	5	4	3	3	7	5
+ 5	+ 6	+ 6	+ 1	+ 7	+ 2	+ 5	+ 4

7	5	7	2	4	5	5	2
+ 1	+ 8	+ 2	+ 5	+ 7	+ 3	+ 8	+ 4

5

A	B	C	D	E
595	924	595	39	37
531	150	103	21	25
326	501	525	53	10
+ 114	+ 531	+ 132	+ 35	+ 50

6

A 3416 + 20 + 512 = ___________

B 17 + 2135 = ___________

Lesson 42

Test	+	Facts	+	Problems	+	Bonus	=	TOTAL

1

$3 + 9$	$7 + 9$	$7 + 5$	$4 + 9$	$5 + 8$	$2 + 9$	$7 + 9$	$8 + 9$
$8 + 5$	$6 + 9$	$8 + 9$	$7 + 5$	$2 + 9$	$6 + 9$	$6 + 5$	$4 + 9$

2

$2 + 7$	$2 + 9$	$2 + 8$	$2 + 6$	$2 + 9$	$2 + 7$	$2 + 8$	$2 + 5$
$2 + 9$	$2 + 7$	$2 + 5$	$2 + 6$	$2 + 8$	$2 + 6$	$2 + 9$	$2 + 7$

3

$3 + 7$	$5 + 7$	$6 + 2$	$5 + 7$	$4 + 7$	$9 + 2$	$6 + 5$	$8 + 2$
$8 + 5$	$7 + 4$	$7 + 2$	$7 + 3$	$5 + 8$	$9 + 2$	$9 + 5$	$5 + 2$

4

A	B	C	D	E	F
$27 + 4$	$13 + 7$	$45 + 8$	$15 + 9$	$27 + 4$	$42 + 8$

5

A ________ B ________ C ________

D ________ E ________ F ________

6

A

The student had a party. He invited 46 boys and 47 girls. He invited 201 parents. He invited 32 teachers. 40 bags of food were ordered. How many people were invited to his party?

B

The machine put 205 chickens on the ship. The machine put 70 apple trees on the ship. The machine put 8 cows on the ship. The machine put 10 horses on the airplane. The machine put 145 rabbits on the ship. How many animals did the machine put on the ship?

C

Some chipmunks stored 27 nuts in a tree. They climbed 38 trees. They stored 42 berries. They stored 5 leaves. They stored 21 more nuts. How many things did the chipmunks store?

Part 6 continues on the next page.

D

John owned 70 donkeys. He sold 17 donkeys. He gave 25 donkeys to his friend. He sold 20 more donkeys. He found 14 squirrels. How many donkeys did John get rid of?

E

The store gave 50 books to the hospital. It gave 80 pens to the team. It gave 20 books to the school. It bought 70 books. It gave 40 books to the library. How many books did the store give away?

F

Chita was going on a long trip. She packed 16 skirts, 18 sweaters, and 37 handkerchiefs. On her trip Chita took pictures of 43 buildings and 32 statues. How many pictures did Chita take on her trip?

A

Yolanda burned 8 pieces of toast. She burned 105 cakes. She burned 64 eggs. Her father burned 23 eggs. Yolanda burned 31 pans of soup. How many things did Yolanda burn?

B

The man put 27 pots in the sink. He put 24 bowls in the sink. The woman put 38 pans in the sink. The man put 33 glasses in the sink. He put 61 cups in the cupboard. How many things did the man put in the sink?

Part 7 continues on the next page.

C

The store cleaned 210 shirts. It cleaned 8 cars. It cleaned 60 pants. It sold 46 parts. It cleaned 120 coats. Add the pieces of clothing the store cleaned.

D

One year Lee's cornfield grew 400 ears of corn. The next year it grew 125 ears of corn. Lee sold 75 pigs. He grew 25 pea plants. How many things grew on Lee's farm?

E

The Travel Light Circus had 70 people working for it. It also had 305 talking birds, 87 dancing bears, and 12 elephants working for it. There were 20 clowns and 20 singers in its show. How many animals worked in the Travel Light Circus?

8

A
$$\begin{array}{r} 3245 \\ 1257 \\ 1723 \\ +\,2134 \\ \hline \end{array}$$

B
$$\begin{array}{r} 1470 \\ 2540 \\ 3430 \\ +\,1220 \\ \hline \end{array}$$

C
$$\begin{array}{r} 1438 \\ 1525 \\ 3440 \\ +\,2322 \\ \hline \end{array}$$

D
$$\begin{array}{r} 3401 \\ 1597 \\ 1255 \\ +\,2145 \\ \hline \end{array}$$

E
$$\begin{array}{r} 2420 \\ 1970 \\ 510 \\ +\,1100 \\ \hline \end{array}$$

A $3425 + 20 =$ ________

B $3748 + 51 + 210 =$ ________

Lesson 43

Facts	+	Problems	+	Bonus	=	TOTAL

1

7 + 6	7 + 5	7 + 7	7 + 5	7 + 7	7 + 6	7 + 5	7 + 6

2

A	B	C	D	E	F
46 + 2	47 + 5	15 + 9	18 + 2	35 + 2	65 + 8

3

5 + 9	7 + 9	4 + 9	8 + 9	6 + 9	8 + 9	7 + 9	4 + 9
2 + 9	3 + 9	8 + 9	4 + 9	2 + 9	3 + 9	2 + 9	8 + 9

4

2 + 9	3 + 7	9 + 5	2 + 5	7 + 5	5 + 7	2 + 7	6 + 5
4 + 7	6 + 5	2 + 9	9 + 5	2 + 6	8 + 5	2 + 8	8 + 5

5

A ________ B ________ C ________

D ________ E ________ F ________

A

The team lost 37 games in May. They tied 18 games in June. They lost 5 games in July. They won 14 games in July. They lost 20 games in August. Add the games the team lost.

B

The store threw away 250 rotten apples. It bought 300 red apples. It gave away 40 dirty apples. It dumped 105 old apples in the garbage. It threw away 200 rotten oranges. How many apples did the store get rid of?

C

Rusty thought he saw a snake. He shivered 11 times. He shook 48 times. He smiled 12 times. He screamed 11 times. How many times did Rusty act scared?

D

Erik called his friend 23 times. Erik rode the bus to his friend's house 15 times. He rode his bike to his friend's house 4 times. He took the bus to school 8 times. He walked to his friend's house 10 times. How many times did he go to his friend's house?

Part 6 continues on the next page.

E

The worker fixed 12 motorcycle tires. He fixed 51 bike tires. He fixed 3 bikes. He fixed 25 truck tires. He fixed 74 trucks. How many tires did the worker fix?

7

A

```
  743
  527
  222
+ 173
-----
```

B

```
  40
  70
  60
+ 20
----
```

C

```
  150
  394
  211
+ 144
-----
```

D

```
  3931
   172
  2492
+  313
------
```

8

A 3000 + 64 + 120 = ________

B 3250 + 70 + 210 = ________

Lesson 44

Facts	+	Problems	+	Bonus	=	TOTAL

1

7	7	7	7	7	7	7	7
+ 8	+ 6	+ 8	+ 7	+ 5	+ 8	+ 6	+ 8

7	7	7	7	7	7	7	7
+ 6	+ 9	+ 8	+ 6	+ 8	+ 7	+ 5	+ 9

2

A	B	C	D	E	F
29	15	27	52	15	59
+ 5	+ 8	+ 2	+ 7	+ 9	+ 2

3

6	4	9	2	5	3	8	6
+ 9	+ 9	+ 9	+ 9	+ 9	+ 9	+ 9	+ 9

4	7	9	7	8	7	2	8
+ 9	+ 9	+ 9	+ 9	+ 9	+ 9	+ 9	+ 9

4

2	8	2	8	9	2	6	4
+ 7	+ 5	+ 6	+ 5	+ 5	+ 9	+ 5	+ 7

2	7	2	9	3	2	5	6
+ 8	+ 5	+ 5	+ 5	+ 7	+ 9	+ 7	+ 5

5

A ________ B ________ C ________ D ________

E ________ F ________

6

A

The machine washed 290 dishes during breakfast. It broke 15 dishes during breakfast. It broke 8 dishes during lunch. The boss broke 5 dishes during lunch. The machine broke 20 dishes during supper. How many dishes did the machine break?

B

The team gave away 48 dirty balls. It threw away 25 old balls. It sent 100 balls to the school. The team bought 300 new balls. The team gave away 15 broken bats. How many balls did the team get rid of?

C

Kirk repaired 122 clocks. Elsa repaired 199 clocks. Walt repaired 39 watches. Walt repaired 195 clocks, and then 115 more clocks. Find out how many clocks Walt repaired.

Part 6 continues on the next page.

D

Leo had 2 cats. He had 4 saws. He had 2 turtles. He had 1 parrot. He wanted to get 4 horses. He had 3 fish. Add the pets that Leo had.

E

The woman had a greenhouse. She bought 470 rosebushes. She was given 38 planting tools. She returned 20 dead plants. She lost 7 watering cans. She sold 25 tulips. She found 210 cans of weed killer. How many things did the woman get?

7

A

$$\begin{array}{r} 370 \\ 920 \\ 310 \\ +\ 190 \\ \hline \end{array}$$

B

$$\begin{array}{r} 590 \\ 4413 \\ 1192 \\ +\ 1710 \\ \hline \end{array}$$

C

$$\begin{array}{r} 2634 \\ 579 \\ 4782 \\ +\ \ 102 \\ \hline \end{array}$$

D

$$\begin{array}{r} 783 \\ 122 \\ 492 \\ +\ 571 \\ \hline \end{array}$$

8

A 2425 + 10 + 20 = ________

B 523 + 4 + 72 = ________

Lesson 45

Facts	+	Problems	+	Bonus	=	TOTAL

1

4	3	2	5	3	4	2	5
+ 4	+ 3	+ 2	+ 5	+ 3	+ 4	+ 2	+ 5

2

A ________

B ________

C ________

D ________

E ________

F ________

3

6	2	4	2	8	5	2	7
+ 9	+ 6	+ 9	+ 8	+ 9	+ 9	+ 7	+ 9

3	8	2	3	9	6	8	5
+ 9	+ 2	+ 9	+ 7	+ 9	+ 5	+ 5	+ 6

4

50	20	40	50	60	10	20
+ 30	+ 40	+ 50	+ 10	+ 20	+ 40	+ 30

5

7	7	7	7	7	7	7	7
+ 9	+ 8	+ 7	+ 6	+ 8	+ 5	+ 8	+ 9

7	7	7	7	7	7	7	7
+ 6	+ 7	+ 8	+ 9	+ 6	+ 8	+ 5	+ 8

6

A ________ **B** ________ **C** ________ **D** ________

E ________ **F** ________

7

A

Diana used 108 nails to build a table. She used 32 pieces of wood to build a bed. She used 45 nails to build a shelf. She used 40 nails to fix a chair. She used 20 nails to build a door. How many nails did Diana use to build things?

B

The short man roasted 35 hamburgers over the campfire. The tall woman roasted 28 marshmallows. The girl roasted 35 marshmallows. The boy roasted 40 marshmallows. The tall man roasted 20 marshmallows. How many marshmallows were roasted?

C

Jan bought 15 comic books. She gave 28 comic books to her mom. She found 10 comic books. She threw away 15 comic books. She gave 10 comic books to the school. How many comic books did Jan get rid of?

Part 7 continues on the next page.

D

373 spiders weren't hungry. 621 children were hungry. 235 skunks were hungry. 315 walruses were dirty. 221 lions were hungry. How many animals were hungry?

E

The store sold 420 big watermelons. It sold 120 red apples. It sold 45 little watermelons. It sold 25 red watermelons. How many watermelons did the store sell?

8

A
$$\begin{array}{r} 3245 \\ 1247 \\ 415 \\ +\ 2101 \\ \hline \end{array}$$

B
$$\begin{array}{r} 443 \\ 472 \\ 468 \\ +\ 515 \\ \hline \end{array}$$

C
$$\begin{array}{r} 95 \\ 42 \\ 59 \\ +\ 11 \\ \hline \end{array}$$

D
$$\begin{array}{r} 4002 \\ 916 \\ 1205 \\ +\ 515 \\ \hline \end{array}$$

9

A $1520 + 31 + 217 =$ ________

B $720 + 2134 + 61 =$ ________

Lesson 46

Test	+	Facts	+	Problems	+	Bonus	=	TOTAL

1

6 + 6	5 + 5	3 + 3	4 + 4	7 + 7	4 + 4	5 + 5	7 + 7
7 + 7	6 + 6	2 + 2	7 + 7	6 + 6	4 + 4	3 + 3	6 + 6

2

A 20 + 50

B 50 + 50

C 70 + 20

D 80 + 50

E 10 + 40

3

A --------

B --------

C --------

D --------

E --------

F --------

4

8 + 9	2 + 8	5 + 9	7 + 9	3 + 9	2 + 7	6 + 9	4 + 9
8 + 5	5 + 6	2 + 9	4 + 5	7 + 5	5 + 9	7 + 4	5 + 7

5

7	7	7	7	7	7	7	7
+ 9	+ 8	+ 7	+ 6	+ 9	+ 5	+ 8	+ 6

7	7	7	7	7	7	7	7
+ 7	+ 6	+ 8	+ 6	+ 5	+ 9	+ 6	+ 8

6

A	B	C	D
1959	2195	1413	3062
1755	717	305	1107
2225	584	2909	559
+ 1121	+ 3502	+ 221	+ 141

7

A --------

B --------

C --------

D --------

E --------

F --------

8

A

39 truck drivers went to work. 11 teachers went to work. 15 bees went to work. 17 carpenters went to lunch. 24 farmers went to work. How many people went to work?

B

Father threw 140 potatoes into his stew. He put in 200 pieces of meat, 50 onions, 9 spices, and 20 carrots. How many vegetables did father put into his stew?

Part 8 continues on the next page.

C

Kate Little Sky earned 27 points during reading. She earned 5 points during math. She earned 20 points during spelling. She used up 40 points during gym. She earned 15 minutes extra recess during history class. How many points did Kate earn?

D

Ramona went to the wax museum 17 times. Tahera went to the art museum 14 times. Amanda went to the wax museum 5 times. Her brother Brian went to the wax museum 10 times. Fritz went to the museum of history 8 times. How many times was the wax museum visited?

E

315 boys were jumping rope. 190 girls were playing hockey. 210 girls were playing tennis. 30 girls were watching a tennis game. 70 girls were playing basketball. How many girls were playing?

A $2041 + 6 + 51 =$ ________

B $87 + 2105 + 420 =$ ________

Lesson 47

Test	+	Facts	+	Problems	+	Bonus	=	TOTAL

1

8 + 8	5 + 5	9 + 9	4 + 4	8 + 8	6 + 6	9 + 9	7 + 7
6 + 6	9 + 9	7 + 7	8 + 8	4 + 4	9 + 9	7 + 7	8 + 8

2

A 90 + 50 = ______ **B** 20 + 60 = ______ **C** 80 + 50 = ______

D 50 + 40 = ______ **E** 70 + 10 = ______

3

A ________ **B** ________ **C** ________ **D** ________

E ________ **F** ________

4

7 + 5	7 + 8	7 + 6	7 + 9	7 + 7	7 + 8	7 + 9	7 + 6
7 + 7	7 + 8	7 + 5	7 + 6	7 + 5	7 + 8	7 + 5	7 + 8

5

3 + 9	2 + 8	6 + 9	4 + 9	2 + 9	2 + 5	8 + 9	7 + 9
7 + 5	5 + 9	4 + 7	6 + 5	3 + 7	9 + 5	8 + 5	7 + 4

6

A ________ B ________ C ________

D ________ E ________ F ________

7

A

$$\begin{array}{r} 2943 \\ 685 \\ 1259 \\ +1102 \\ \hline \end{array}$$

B

$$\begin{array}{r} 1812 \\ 2128 \\ 1724 \\ +2205 \\ \hline \end{array}$$

C

$$\begin{array}{r} 435 \\ 752 \\ 219 \\ +201 \\ \hline \end{array}$$

D

$$\begin{array}{r} 9 \\ 41 \\ 60 \\ +18 \\ \hline \end{array}$$

8

A	5001	4064	2505	340
B	9052	7008	520	3706
C	980	4506	7009	4031
D	6102	870	9003	6041
E	750	8002	5013	4106
F	8309	260	5004	3026

9

A $85 + 130 + 4021 =$ ________

B $85 + 1423 + 210 =$ ________

10

A

The musician owned 60 violins. He owned 3 grand pianos. He played in 248 live concerts during the year. He owned 5 guitars and 20 trumpets. How many instruments did he own?

B

Vera beat her drums 38 times. She snapped her fingers 19 times. Her brother snapped his fingers 21 times. Vera stamped her feet 63 times. She snapped her fingers 4 times. How many times did Vera snap her fingers?

C

123 rabbits sat in the shade. 235 bears sat in the water. 745 cows sat in the shade. 95 hunters sat in the shade. 30 horses sat in the shade. How many animals sat in the shade?

D

Lamar drew 8 pictures. He threw away 9 pictures. He sent 11 pictures to his grandmother. He gave 5 pictures to his friend. His mom gave Lamar 8 pictures. Add the pictures Lamar got rid of.

E

Asha typed 253 business letters. Grant addressed 272 envelopes. He put stamps on 112 envelopes. Debra answered the phone 151 times. Angus mailed 11 envelopes. How many letters and envelopes were handled?

Lesson 48

Facts	+	Problems	+	Bonus	=	TOTAL

1

9	6	7	4	8	6	9	7
+ 9	+ 6	+ 7	+ 4	+ 8	+ 6	+ 9	+ 7

8	4	9	8	9	6	9	8
+ 8	+ 4	+ 9	+ 8	+ 9	+ 6	+ 9	+ 8

2

A 70 + 20 = ______ **B** 50 + 70 = ______ **C** 40 + 20 = ______

D 80 + 20 = ______ **E** 40 + 50 = ______

3

A ________ **B** ________ **C** ________ **D** ________

4

A ☐ { 6 __________________ ; 9 __________________

B ☐ { 8 __________________ ; 9 __________________

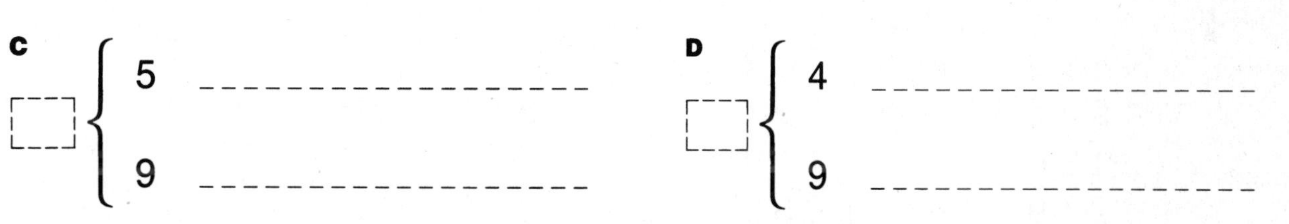

C ☐ { 5 __________________ ; 9 __________________

D ☐ { 4 __________________ ; 9 __________________

E ☐ { 3 __________________ ; 9 __________________

F ☐ { 7 __________________ ; 9 __________________

5

7	4	7	8	7	6	7	3
+ 9	+ 9	+ 6	+ 9	+ 8	+ 9	+ 5	+ 9

7	7	5	7	7	7	2	7
+ 8	+ 5	+ 9	+ 7	+ 6	+ 9	+ 5	+ 8

6

A 34 + 4 = ______ B 15 + 7 = ______ C 12 + 6 = ______

D 37 + 5 = ______ E 32 + 4 = ______ F 17 + 5 = ______

G 12 + 4 = ______ H 43 + 5 = ______ I 28 + 2 = ______

J 76 + 5 = ______

7

Add each number to 11.

4	6	3	1	0	8	2	5
☐	☐	☐	☐	☐	☐	☐	☐

8

A	B	C	D
4490	732	492	582
1225	297	522	119
2130	119	581	514
+ 1219	+ 652	+ 312	+ 143

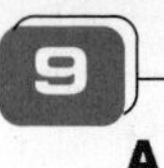

A

38 teachers liked to exercise. 24 eye doctors liked to exercise. 45 foot doctors liked to read books. 39 brain doctors liked to exercise. 25 foot doctors liked to exercise. How many doctors liked to exercise?

B

Lighthouse A blinked 240 times. Lighthouse B blinked 120 times. Lighthouse C blinked 150 times. 50 ships passed lighthouse A. 10 ships passed lighthouse B. 210 ships passed lighthouse C. How many times did lighthouse B and lighthouse C blink?

C

The shoemaker made 8 pairs of boots. He repaired 32 pairs of shoes. He made 105 pairs of shoes. He made 21 leather purses. How many things did the shoemaker make?

D

10 wolves lived on the grounds of a science laboratory. 34 bears lived in the woods. 52 alligators lived in a swamp. 16 raccoons lived in the woods. 24 deer lived in the woods. How many animals lived in the woods?

E

The workers cleaned 493 machines. They cleaned 222 floors. They fixed 80 cars. They cleaned 134 trucks. How many things did the workers clean?

Lesson 49

Facts	+	Problems	+	Bonus	=	TOTAL

1

9 + 6	9 + 5	9 + 7	9 + 3	9 + 8	9 + 4	9 + 8	9 + 2
9 + 4	9 + 7	9 + 6	9 + 5	9 + 7	9 + 8	9 + 4	9 + 6

2

A 60 + 90 = ______ **B** 40 + 20 = ______ **C** 20 + 30 = ______

D 70 + 30 = ______ **E** 70 + 70 = ______

3

A ________ **B** ________ **C** ________ **D** ________

4

9 + 9	7 + 7	8 + 8	9 + 9	6 + 6	7 + 7	8 + 8	7 + 7
6 + 6	5 + 5	9 + 9	8 + 8	6 + 6	7 + 7	9 + 9	8 + 8

5

7 + 8	6 + 9	7 + 5	4 + 9	3 + 9	8 + 9	7 + 6	7 + 9
7 + 7	7 + 6	7 + 9	2 + 5	7 + 8	7 + 5	5 + 9	7 + 8

6

A 18 + 5 = ______ B 14 + 5 = ______ C 42 + 7 = ______

D 58 + 2 = ______ E 31 + 6 = ______ F 14 + 6 = ______

G 42 + 5 = ______ H 17 + 4 = ______ I 23 + 3 = ______

J 13 + 7 = ______

7

Add each number to 12.

3	6	2	4	5	1	7	0

8

A
$$\begin{array}{r} 843 \\ 172 \\ 749 \\ +3133 \\ \hline \end{array}$$

B
$$\begin{array}{r} 623 \\ 114 \\ 126 \\ +140 \\ \hline \end{array}$$

C
$$\begin{array}{r} 327 \\ 219 \\ 423 \\ +111 \\ \hline \end{array}$$

D
$$\begin{array}{r} 364 \\ 169 \\ 1223 \\ +\ \ 922 \\ \hline \end{array}$$

9

A

275 men chopped wood. 86 students worked math problems. 42 women chopped wood. 36 men chopped onions. 181 girls chopped wood. Add the people who chopped wood.

Part 9 continues on the next page.

B

The store sold 91 electric trains. It threw away 59 broken games. It sold 97 puzzles. It sold 11 walking dolls. It sold 11 magic sets. It ordered 23 stuffed animals for its shelves. Add the things the store sold.

C

Elaine sharpened 451 crayons. She used 81 pencils. She sharpened 270 saws. She sharpened 161 scissors. Doug sharpened 215 pencils. Elaine sharpened 116 pencils. How many objects did Elaine sharpen?

D

In the Delicious Food Restaurant, there are 8 boys who clean tables. There are 25 girls who wait on tables. There are 25 boys who wait on tables. 18 people order something to eat. Add the people who work at the restaurant.

E

Carla studied math for 90 minutes. Rico studied spelling for 50 minutes. Carla took a nap for 20 minutes. Carla studied spelling for 50 minutes. Carla studied science for 200 minutes. How many minutes did Carla study?

Lesson 50

Facts + Problems + Bonus = TOTAL

1

9	9	9	9	9	9	9	9
+ 6	+ 4	+ 9	+ 3	+ 7	+ 2	+ 5	+ 8

9	9	9	9	9	9	9	9
+ 5	+ 7	+ 2	+ 8	+ 6	+ 4	+ 9	+ 3

2

A 30 + 90 = ______ **B** 70 + 70 = ______ **C** 60 + 20 = ______

D 40 + 50 = ______ **E** 90 + 10 = ______

3

A ________ **B** ________ **C** ________

D ________ **E** ________

4

A ☐ { 7 ________________ / 8 ________________

B ☐ { 7 ________________ / 6 ________________

C ☐ { 7 ________________ / 5 ________________

D ☐ { 7 ________________ / 9 ________________

5

6 + 6	7 + 5	8 + 8	7 + 8	9 + 9	7 + 6	7 + 7	7 + 9
9 + 9	4 + 9	7 + 9	3 + 9	6 + 6	6 + 9	8 + 8	7 + 7

6

A 15 + 9 = ______

B 45 + 4 = ______

C 15 + 3 = ______

D 43 + 5 = ______

E 25 + 9 = ______

F 17 + 2 = ______

G 17 + 9 = ______

H 11 + 9 = ______

I 47 + 5 = ______

J 24 + 4 = ______

7

Add each number to 12.

4	2	6	3	1	0	7	5

8

A

$$\begin{array}{r} 1787 \\ 2155 \\ 1146 \\ +\ 1201 \\ \hline \end{array}$$

B

$$\begin{array}{r} 3217 \\ 1681 \\ 531 \\ +\ 1455 \\ \hline \end{array}$$

C

$$\begin{array}{r} 1098 \\ 2755 \\ 1545 \\ +\ 1311 \\ \hline \end{array}$$

D

$$\begin{array}{r} 461 \\ 75 \\ 126 \\ +\ \ 23 \\ \hline \end{array}$$

A

An owl fell asleep 18 times. He hunted 5 times. He blinked his eyes 16 times. He got materials to build his nest 22 times. How many times did the owl work?

B

Casey saddled 29 horses. He rode 125 horses through the fields. He brushed 3 horses. He fed sugar to 9 horses. He rode 14 horses through the woods. How many horses did Casey ride?

C

The store ordered 180 shirts. It ordered 3 axes. It ordered 1450 hats. It sold 45 hats. It ordered 7 shirts. Add the clothes the store ordered.

D

The boy folded 192 newspapers. He folded 12 napkins. He folded 65 newspapers. His grandfather folded 28 newspapers. His father folded 56 newspapers. The boy folded 4 newspapers. How many newspapers did the boy fold?

E

The man owned a wig factory. He made 92 wigs the first day. He sold 145 wigs the next day. He placed ads in 18 magazines. Next day he sold 52 wigs and sent out 63 bills to customers. Later he made 211 more wigs. How many wigs did the man make?

F

Russ put 32 cans of beans into the grocery cart. Carrie put 28 bars of soap into the cart. May put in 15 bottles of pop. 17 jars of honey fell out of the cart. Abner put 11 turkeys into the cart. How many things made the cart heavier?

Lesson 51

Test + Facts + Problems + Bonus = TOTAL

1

8	6	9	5	6	8	8	9
+ 7	+ 7	+ 7	+ 7	+ 7	+ 7	+ 7	+ 7

8	6	7	8	6	8	7	8
+ 7	+ 7	+ 7	+ 7	+ 7	+ 7	+ 7	+ 7

2

A 60 + 60 = ______ **B** 60 + 20 = ______ **C** 70 + 70 = ______

D 20 + 30 = ______ **E** 80 + 20 = ______

3

A

9	9	9	9	9	9	9	9
+ 5	+ 4	+ 7	+ 3	+ 8	+ 5	+ 6	+ 8

B

9	9	9	9	9	9	9	9
+ 3	+ 9	+ 7	+ 6	+ 2	+ 8	+ 6	+ 7

4

A 45 + 8 = ______ **B** 12 + 7 = ______ **C** 17 + 6 = ______

D 14 + 3 = ______ **E** 27 + 5 = ______ **F** 62 + 5 = ______

G 47 + 3 = ______ **H** 18 + 5 = ______ **I** 14 + 9 = ______

J 31 + 7 = ______

5

7 + 6	4 + 7	4 + 4	3 + 9	5 + 4	2 + 6	6 + 9	7 + 5
8 + 7	3 + 3	8 + 9	3 + 5	2 + 9	9 + 7	9 + 5	8 + 8
6 + 6	7 + 9	3 + 7	8 + 5	9 + 9	2 + 2	5 + 9	2 + 8
4 + 9	2 + 7	5 + 5	7 + 7	5 + 7	2 + 9	5 + 8	7 + 4

6

A ________ B ________ C ________ D ________

E ________ F ________

7

Add each number to 15.

5	3	7	2	8	4	9	1

8

A	B	C	D
908	1537	214	1109
4865	675	167	1802
153	2253	786	2494
+ 1130	+ 420	+ 520	+ 1484

A

Art was watching television. He dozed off for 9 minutes. He woke up for 4 minutes. He dozed off for 7 minutes. He woke up for a moment, then dozed off again for 5 more minutes. How many minutes did Art doze off?

B

Kermit made salads for a hotel restaurant. Monday he made 20 salads. Tuesday was his day off and he baked 30 cookies. Wednesday he made 40 salads. After work he took 10 shirts to the cleaners. Thursday he made 10 salads; Friday he made 50 salads. Add the number of salads Kermit made that week.

C

Faye got a new camera. She took 14 pictures. She developed 82 pictures. She pasted 20 pictures in her photo book. She developed 426 pictures and gave away 52 pictures. She developed 28 more pictures. Add the pictures that Faye developed.

D

The hotel sold 40 old chairs. It bought 135 plastic chairs. It gave away 35 broken chairs. It bought 2040 wooden chairs. It bought 50 plastic tables. It bought 5 steel chairs. How many chairs did the hotel buy?

E

375 oak trees were hit by lightning. 108 schools were hit by lightning. 104 apartment houses were hit by lightning. 85 apartment houses escaped the lightning. How many buildings were hit by lightning?

Lesson 52

Facts	+	Problems	+	Bonus	=	TOTAL

1

$9 + 6$ | $9 + 4$ | $9 + 9$ | $9 + 7$ | $9 + 5$ | $9 + 9$ | $9 + 8$ | $9 + 7$

$9 + 2$ | $9 + 7$ | $9 + 3$ | $9 + 6$ | $9 + 2$ | $9 + 6$ | $9 + 3$ | $9 + 8$

2

A 90 + 90 = ______ **B** 30 + 50 = ______ **C** 60 + 10 = ______

D 70 + 70 = ______ **E** 80 + 50 = ______

3

$10 + 7$ | $5 + 7$ | $8 + 7$ | $6 + 7$ | $5 + 7$ | $8 + 7$ | $6 + 7$ | $8 + 7$

$7 + 7$ | $9 + 7$ | $6 + 7$ | $8 + 7$ | $5 + 7$ | $6 + 7$ | $8 + 7$ | $6 + 7$

4

$9 + 9$ | $5 + 9$ | $8 + 8$ | $4 + 9$ | $8 + 9$ | $7 + 7$ | $5 + 9$ | $5 + 5$

$7 + 5$ | $6 + 6$ | $7 + 8$ | $7 + 7$ | $8 + 8$ | $7 + 6$ | $2 + 9$ | $9 + 9$

5

A 17 + 7 = ______ B 14 + 5 = ______ C 19 + 5 = ______

D 34 + 4 = ______ E 47 + 5 = ______ F 12 + 8 = ______

G 15 + 3 = ______ H 52 + 6 = ______ I 37 + 8 = ______

J 24 + 3 = ______

6

A ________ B ________ C ________ D ________

E ________ F ________

7

Add each number to 15.

2	7	3	5	8	1	6	9

8

A
$$\begin{array}{r} 3042 \\ 553 \\ 224 \\ +\,2142 \\ \hline \end{array}$$

B
$$\begin{array}{r} 83 \\ 54 \\ 47 \\ +\,12 \\ \hline \end{array}$$

C
$$\begin{array}{r} 3147 \\ 185 \\ 124 \\ +\,2035 \\ \hline \end{array}$$

D
$$\begin{array}{r} 1924 \\ 231 \\ 465 \\ +\,3133 \\ \hline \end{array}$$

A

Mr. Hawk drove his car to the city 96 times. He thought of the city 14 times. He ran to the city 45 times. He walked to the store 23 times. He rode his bike to the city 57 times. How many times did Mr. Hawk go to the city?

B

The company sent 238 books to England. It sent 418 magazines to Canada. It sent 4235 magazines to France. It got 524 magazines from China. It sent 150 magazines to Germany. How many magazines did the company send?

C

284 people sat in the park. 28 people sat on their porch. 9 people paid bills. 35 cleaned house. 43 people washed their car.
How many people did things?

D

Connie gave 74 bells to her brother. She bought 35 bells from the store. She threw 20 bells and 8 comic books in the garbage. She dumped 6 bells in the river. How many bells did she get rid of?

E

The machine crushed 314 bags of ice. It broke down for 25 minutes. Then it crushed 3104 bags of ice. Later it crushed 31 more bags of ice. It broke down again for 28 minutes. Add the bags of ice the machine crushed.

Lesson 53

Facts	+	Problems	+	Bonus	=	TOTAL

1

$6 + 3$	$6 + 1$	$6 + 4$	$6 + 5$	$6 + 4$	$6 + 3$	$6 + 5$	$6 + 3$
$6 + 5$	$6 + 3$	$6 + 4$	$6 + 2$	$6 + 5$	$6 + 3$	$6 + 4$	$6 + 3$

2

A $80 + 80 =$ ______ **B** $30 + 30 =$ ______ **C** $70 + 50 =$ ______

D $10 + 60 =$ ______ **E** $50 + 50 =$ ______

3

$5 + 7$	$9 + 7$	$6 + 7$	$7 + 7$	$8 + 7$	$6 + 7$	$8 + 7$	$6 + 7$
$9 + 7$	$8 + 7$	$6 + 7$	$8 + 7$	$5 + 7$	$9 + 7$	$8 + 7$	$6 + 7$

4

$9 + 5$	$6 + 6$	$9 + 8$	$7 + 7$	$9 + 4$	$9 + 7$	$8 + 8$	$9 + 2$
$9 + 9$	$3 + 9$	$9 + 6$	$7 + 8$	$9 + 3$	$7 + 6$	$4 + 9$	$9 + 9$

5

A 29 + 5 = ______ B 19 + 7 = ______ C 12 + 6 = ______

D 42 + 7 = ______ E 13 + 9 = ______ F 14 + 7 = ______

G 43 + 5 = ______ H 15 + 9 = ______ I 53 + 7 = ______

J 31 + 6 = ______

6

A ________ B ________ C ________ D ________

E ________ F ________

7

Add each number to 12.

2	8	4	3	9	6	7	5

8

A
```
  2831
  1540
   490
+ 4127
```

B
```
   185
  2764
   444
+ 1315
```

C
```
  1821
  1374
  2355
+ 1240
```

D
```
  3027
   332
   653
+  434
```

A

The woman wrote 4598 greeting cards. She wrote 192 poems. She had 365 books in her library at home. 43 of them were mysteries. She wrote 1214 short stories. How many things did the woman write?

B

There were 86 chickens in the old barn. There were 24 horses in the new barn. There were 7 cows in the new barn. There were 341 rabbits in the new barn. There were 8 tractors in the new barn. How many animals were in the new barn?

C

Margarita found 205 bottle caps in the park. She sold 30 bottle caps to her friend. She found 90 paper cups in the park. She bought 80 bottle caps. Her sister gave her 15 bottle caps. How many bottle caps did Margarita get?

D

The company sold 200 pickup trucks. The company bought 80 dump trucks. It bought 50 pickup trucks. It bought 90 new cars. It bought 40 bikes. How many trucks did the company buy?

E

Rick counted 38 telephone poles. He counted 142 cows. He counted 450 horses. He counted 3410 sheep. Helen counted 200 deer. How many animals did Rick count?

Lesson 54

Facts + Problems + Bonus = TOTAL

1

6 + 4	6 + 1	6 + 3	6 + 5	6 + 2	6 + 5	6 + 4	6 + 3
6 + 5	6 + 4	6 + 3	6 + 5	6 + 2	6 + 4	6 + 3	6 + 5

2

A 80 + 50 = ______ **B** 10 + 70 = ______ **C** 70 + 70 = ______

D 90 + 90 = ______ **E** 30 + 30 = ______

3

5 + 7	6 + 7	9 + 7	8 + 7	6 + 7	9 + 7	6 + 7	8 + 7
9 + 7	5 + 7	8 + 7	6 + 7	5 + 7	8 + 7	7 + 7	6 + 7

4

9 + 8	7 + 7	9 + 3	9 + 7	8 + 8	9 + 4	9 + 9	9 + 6
6 + 6	7 + 8	7 + 5	3 + 3	9 + 8	7 + 6	9 + 5	2 + 9

5

A $17 + 8 =$ ______ B $14 + 3 =$ ______ C $38 + 5 =$ ______

D $25 + 4 =$ ______ E $12 + 5 =$ ______ F $14 + 7 =$ ______

G $34 + 4 =$ ______ H $46 + 6 =$ ______ I $18 + 8 =$ ______

J $19 + 6 =$ ______

6

A ________ B ________ C ________ D ________

7

Add each number to 15.

4	2	8	3	6	7	9	5

8

A
$$\begin{array}{r} 2177 \\ 1358 \\ 3542 \\ +\,1727 \\ \hline \end{array}$$

B
$$\begin{array}{r} 178 \\ 298 \\ 154 \\ +\,525 \\ \hline \end{array}$$

C
$$\begin{array}{r} 3014 \\ 964 \\ 1958 \\ +\quad 643 \\ \hline \end{array}$$

D
$$\begin{array}{r} 248 \\ 568 \\ 942 \\ +\,432 \\ \hline \end{array}$$

A

50 Girl Scouts got in the bus. A man put 50 suitcases in the bus. 80 Boy Scouts got in the bus. 80 people got out of the bus. The man put 60 boxes in the car. The man took 20 boxes out of the bus. How many things made the bus heavier?

B

499 cars were made of metal. 4192 tools were made of metal. 621 tools were made of plastic. 374 houses were made of wood. 13 clothing trunks were made of metal. How many objects were made of metal?

C

The Buffalo Marching Band marched at 137 games. They played music during 168 games. They toured 63 cities with the team. They marched at 12 games. Their team played 8 more games. The band marched during 4 of them. How many games did the band march in?

D

35 party guests went upstairs. 24 party helpers went downstairs. 319 more guests went upstairs. 2 neighbors went upstairs. 21 guests went outdoors. How many people went upstairs?

E

In the morning Pierre gave French lessons to 2 students. In the afternoon he went sailing with 18 friends. That night he taught French to 53 students. There were 3 other teachers in his school. On Saturdays, Pierre taught French to 101 students. How many students did Pierre have in all?

Lesson 55

Test	+	Facts	+	Problems	+	Bonus	=	TOTAL

1

4	4	4	4	4	4	4	4
+ 5	+ 4	+ 7	+ 6	+ 3	+ 7	+ 5	+ 6

4	4	4	4	4	4	4	4
+ 7	+ 3	+ 5	+ 7	+ 6	+ 7	+ 3	+ 6

2

A 80 + 80 = ______ **B** 70 + 50 = ______ **C** 30 + 10 = _____

D 90 + 10 = ______ **E** 20 + 70 = ______

3

6	6	6	6	6	6	6	6
+ 1	+ 5	+ 3	+ 6	+ 5	+ 4	+ 3	+ 4

6	6	6	6	6	6	6	6
+ 3	+ 5	+ 4	+ 3	+ 5	+ 2	+ 4	+ 6

4

A 14 + 7 = ______ **B** 13 + 5 = ______ **C** 27 + 5 = ______

D 42 + 6 = ______ **E** 13 + 7 = ______ **F** 16 + 3 = ______

G 48 + 8 = ______ **H** 15 + 9 = ______ **I** 34 + 4 = ______

J 23 + 5 = ______

5

7	9	8	8	9	6	6	7
+ 9	+ 3	+ 8	+ 7	+ 6	+ 7	+ 6	+ 6

8	9	9	9	5	9	7	8
+ 7	+ 8	+ 7	+ 4	+ 9	+ 9	+ 6	+ 7

6

A ________ **B** ________ **C** ________ **D** ________

7

A ________ **B** ________ **C** ________ **D** ________

E ________

8

A	B	C	D
2471	574	463	2777
1509	382	573	1578
1409	941	340	1352
+ 1701	+ 111	+ 186	+ 3525

9

A

Clark took care of babies. He sat with baby Jones for 45 minutes. He sat with baby Smith for 200 minutes. He watched TV for 15 minutes. He read for 15 minutes. Then he looked after baby Montoya for 10 minutes. How many minutes did Clark take care of babies?

Part 9 continues on the next page.

B

Rudy had a juice stand. He sold 27 glasses of orange juice and 24 glasses of lemonade. He bought 49 boxes of oranges and 8 boxes of lemons. He threw out 17 rotten carrots. Next Rudy sold 24 glasses of coconut juice and 47 glasses of carrot juice. He cracked open 34 coconuts. How many drinks did Rudy sell?

C

The men and women drove bulldozers for 422 hours. They moved broken tree branches for 874 hours. They laid new lawn for 111 hours. They laid brick for 262 hours. They rested for 433 hours. How many hours did the men and women work?

D

Ken bought 12 bottles. He broke 4 bottles. He sold 16 bottles. He found 40 bottles. He bought 2 jackets. He bought 35 bottles. How many bottles did Ken get?

E

Marta found 314 buttons in a box. She sold 78 buttons to a friend. Her grandmother gave her 20 more buttons. She bought 100 buttons. Her brother gave her 5000 buttons. How many buttons did Marta get?

Lesson 56

Facts	+	Problems	+	Bonus	=	TOTAL

1

4 + 7	4 + 2	4 + 6	4 + 3	4 + 7	4 + 5	4 + 7	4 + 6
4 + 5	4 + 4	4 + 7	4 + 3	4 + 6	4 + 5	4 + 7	4 + 6

2

A 700 + 500 = ________ **B** 300 + 300 = ________

C 600 + 600 = ________ **D** 800 + 100 = ________

3

A [ ] { 6 ________________ ; 3 ________________ }

B [ ] { 6 ________________ ; 5 ________________ }

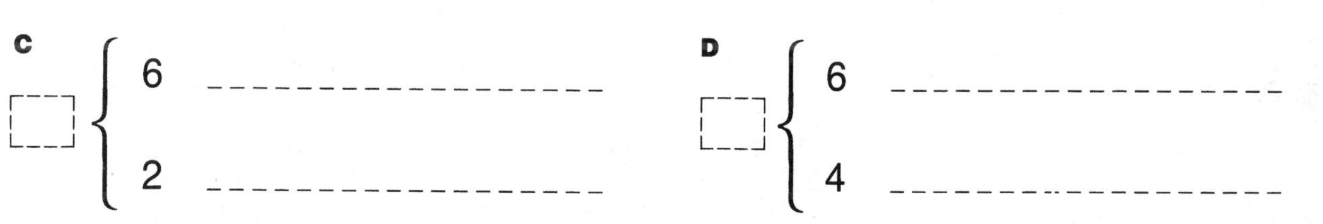

4

5 + 7	9 + 7	9 + 9	7 + 7	6 + 7	5 + 9	8 + 7	8 + 8
7 + 8	9 + 6	7 + 5	7 + 8	6 + 6	9 + 8	9 + 9	7 + 6

A 14 + 7 = ______ **B** 12 + 6 = ______ **C** 38 + 7 = ______

D 32 + 5 = ______ **E** 15 + 4 = ______ **F** 19 + 6 = ______

G 46 + 6 = ______ **H** 42 + 7 = ______ **I** 18 + 8 = ______

J 53 + 7 = ______

6

A ________ **B** ________ **C** ________ **D** ________

E ________ **F** ________

7

Add each number to 17.

1	8	3	5	2	0	7	9

8

A	B	C	D
573	309	754	560
4285	548	683	2420
958	121	339	580
+ 1235	+ 101	+ 656	+ 4090

A

Salvador made 130 jars of applesauce in 1998. He made 150 jars of applesauce in 1999. He made 120 jars of applesauce in 2000. Jackie made 160 jars of applesauce in 2000. How many jars of applesauce did Salvador make?

B

Sheila counted 95 birds in the sky. She counted 452 fishing nets. She counted 47 fishing boats. She counted 22 fishing poles. Cora counted 8 boats. How many fishing things did Sheila count?

C

There are 528 apple trees. There are 4108 cherry trees. There are 180 rose bushes. There are 2400 pear trees. There are 600 oak trees. How many trees are there?

D

Faye bought 384 shares of ABC stock. She bought 2400 shares of DEF stock. She sold 1500 shares of GHI stock. She bought 56 shares of JKL stock. Her brother bought 24 boxes of soap powder. How many shares of stock did Faye get?

E

The machine made 1048 red pencils. It broke 84 green pencils. It made 3400 red rulers. It made 3200 blue pencils. It made 84 pink pencils. How many pencils did the machine make?

Lesson 57

Facts	+	Problems	+	Bonus	=	TOTAL

1

4	4	4	4	4	4	4	4
+ 7	+ 6	+ 4	+ 7	+ 5	+ 7	+ 3	+ 6

4	4	4	4	4	4	4	4
+ 7	+ 5	+ 6	+ 7	+ 4	+ 6	+ 5	+ 7

2

A 800 + 800 = ________ **B** 60 + 20 = ________

C 70 + 70 = ________ **D** 900 + 300 = ________

3

5	3	4	6	2	5	3	6
+ 6	+ 6	+ 6	+ 6	+ 6	+ 6	+ 6	+ 6

4	5	6	3	4	3	5	4
+ 6	+ 6	+ 6	+ 6	+ 6	+ 6	+ 6	+ 6

4

9	6	9	6	7	8	3	8
+ 7	+ 6	+ 4	+ 7	+ 7	+ 7	+ 9	+ 9

7	8	7	9	6	7	7	4
+ 5	+ 8	+ 9	+ 2	+ 9	+ 6	+ 8	+ 9

5

A 18 + 7 = ______

B 15 + 4 = ______

C 27 + 7 = ______

D 41 + 8 = ______

E 12 + 7 = ______

F 12 + 9 = ______

G 39 + 9 = ______

H 23 + 5 = ______

I 15 + 9 = ______

J 27 + 4 = ______

6

A ________

B ________

C ________

D ________

E ________

F ________

7

Add each number to 17.

9	7	4	8	0	3	6	5

8

A

```
  1392
  2146
   197
+ 3028
------
```

B

```
  4567
  1707
   505
+ 1716
------
```

C

```
  78
  99
+ 95
----
```

D

```
  5938
  1698
  1801
+  457
------
```

A

The company hired 685 carpenters. It hired 439 truck drivers. It sent bills to 263 people. It hired 517 secretaries. It bought 308 trucks. How many people did the company hire?

B

Shoppers bought 241 beef roasts. They bought 352 beets. They looked at 193 beans. They bought 528 carrots. How many vegetables did shoppers buy?

C

1425 skydivers jumped out of planes. 842 of them landed in the ocean. 84 skydivers jumped out of a plane. 32 skydivers landed in the woods. How many skydivers jumped out of planes?

D

Coley served 24 plates of spaghetti at dinner. He cut 18 slices of bread. He put 19 bowls of spaghetti sauce on the table. Abby put 28 plates of vegetables on the table. 11 more plates of spaghetti were served. 11 glasses of milk were drunk. Add the things that contained spaghetti or spaghetti sauce.

E

Margarita taught 345 lessons in April. She wrote 29 lessons in May. She taught 85 lessons in June. Tomas taught 200 lessons in July. Margarita taught 9 lessons in August. How many lessons did Margarita teach?

Lesson 58

Facts + Problems + Bonus = TOTAL

1

3	5	4	2	3	5	6	4
+ 6	+ 6	+ 6	+ 6	+ 6	+ 6	+ 6	+ 6

5	6	5	3	4	5	3	4
+ 6	+ 6	+ 6	+ 6	+ 6	+ 6	+ 6	+ 6

2

A 50 + 40 = ________ **B** 70 + 50 = ________

C 300 + 500 = ________ **D** 60 + 60 = ________

E 800 + 800 = ________

3

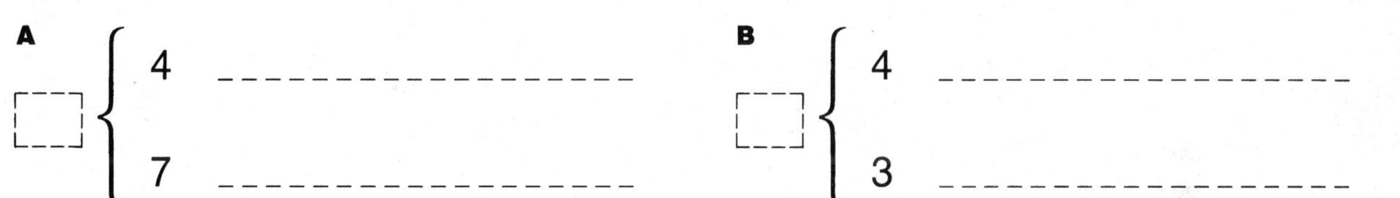

A { 4 ________ / 7 ________ }

B { 4 ________ / 3 ________ }

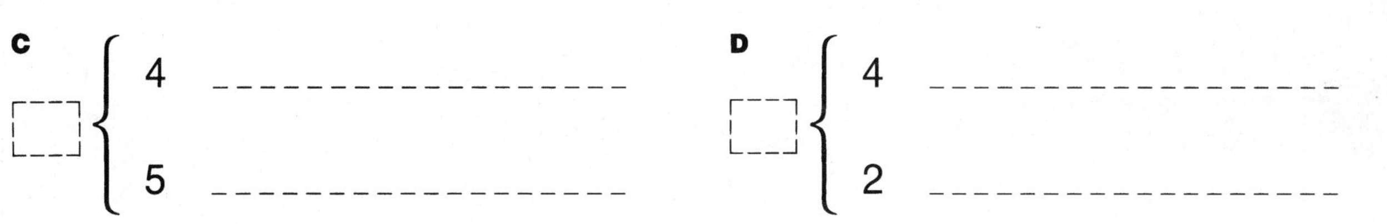

C { 4 ________ / 5 ________ }

D { 4 ________ / 2 ________ }

E { 4 ________ / 1 ________ }

F { 4 ________ / 6 ________ }

4

4	9	4	5	8	9	4	6
+ 3	+ 7	+ 5	+ 7	+ 7	+ 6	+ 6	+ 7

9	4	2	4	8	4	6	9
+ 9	+ 7	+ 9	+ 6	+ 8	+ 9	+ 6	+ 3

5

A 17 + 7 = ______ B 14 + 4 = ______ C 32 + 8 = ______

D 73 + 5 = ______ E 12 + 5 = ______ F 38 + 8 = ______

G 19 + 6 = ______ H 42 + 7 = ______ I 18 + 7 = ______

J 51 + 6 = ______

6

A ________ B ________ C ________ D ________

E ________ F ________

7

Add each number to 14.

3	8	2	5	9	4	7	6
☐	☐	☐	☐	☐	☐	☐	☐

8

A	B	C	D
723	3014	1658	2439
862	89	895	819
287	45	2924	1725
+ 566	+ 1037	+ 3307	+ 494

A

126 snakes crawled. 244 bugs crawled. 178 worms crawled. 196 babies crawled. 57 bugs hopped. 38 bugs crawled. How many animals crawled?

B

Box A had 420 letters in it. Box B had 415 letters in it. Box A had 2140 magazines and 320 cards in it. Box B had 80 newspapers in it. How many things were in Box A?

C

The Rams won 6 rounds in the spelling contest. The Cowboys won 12 rounds. The Jets and the Cowboys tied 3 times. Then the Jets lost 4 rounds. The Rams lost 3 rounds, then won 15 rounds. How many rounds in the spelling contest did the Rams win?

D

The eagle flew over 1600 trees. The eagle flew over 900 houses. The owl flew over 84 roads. The eagle flew over 150 cars. The eagle flew over 470 hills. The eagle sat in 4 treetops. Add the things the eagle flew over.

E

The car passed 96 runners. Then the car passed 391 trucks. It stopped at 8 rest areas. Then it passed 450 trucks. Then the car passed 189 more trucks. Find out how many trucks the car passed.

Lesson 59

Facts	+	Problems	+	Bonus	=	TOTAL

1

8 + 5	8 + 1	8 + 3	8 + 2	8 + 4	8 + 5	8 + 5	8 + 4
8 + 4	8 + 3	8 + 2	8 + 5	8 + 4	8 + 3	8 + 4	8 + 5

2

A 60 + 40 = ________ **B** 900 + 500 = ________

C 80 + 80 = ________ **D** 40 + 30 = ________

E 300 + 500 = ________

3

3 + 4	5 + 4	7 + 4	6 + 4	3 + 4	7 + 4	5 + 4	6 + 4
7 + 4	6 + 4	3 + 4	2 + 4	6 + 4	5 + 4	7 + 4	4 + 4

4

3 + 6	4 + 3	5 + 6	4 + 7	7 + 9	4 + 6	5 + 7	7 + 8
9 + 4	9 + 6	8 + 8	3 + 9	7 + 8	9 + 9	5 + 9	7 + 6

5

A 18 + 8 = ______ B 14 + 3 = ______ C 27 + 7 = ______

D 39 + 9 = ______ E 13 + 6 = ______ F 49 + 7 = ______

G 15 + 8 = ______ H 53 + 4 = ______ I 16 + 5 = ______

J 43 + 5 = ______

6

A ________ B ________ C ________ D ________

E ________ F ________

7

Add each number to 15.

1	9	3	7	8	6	0	5

8

A

$$\begin{array}{r} 964 \\ 672 \\ 632 \\ +361 \\ \hline \end{array}$$

B

$$\begin{array}{r} 569 \\ 987 \\ 234 \\ +759 \\ \hline \end{array}$$

C

$$\begin{array}{r} 347 \\ 476 \\ 29 \\ +141 \\ \hline \end{array}$$

D

$$\begin{array}{r} 838 \\ 298 \\ 631 \\ +620 \\ \hline \end{array}$$

A

93 railroad cars carried coal. 1624 railroad cars carried wood. 538 boats carried steel. 4587 trucks carried coal. 716 boats carried coal. Add the things that carried coal.

B

Craig delivered 2 bags of groceries to Mr. Valdez. He delivered 5 bags of groceries to Mrs. Turner. He loaded 6 bags of groceries onto his truck. Craig delivered 9 bags to Mrs. Ito and 4 bags to Mrs. Brown. He stopped for gas 7 times. He delivered 2 bags to Mr. Fink. How many bags did Craig deliver?

C

Lightning started 31 forest fires. Very dry, hot weather started 28 forest fires. Lightning started 59 forest fires. Fire fighters put out 35 forest fires. How many forest fires did lightning start?

D

The school had a Bake Sale. It sold 234 cakes the first hour. It sold 14 boxes of cookies. It sold 125 cakes the second hour. It sold 84 cakes the third hour. How many cakes did the school sell?

E

Susana went on a game show. She earned 80 points during one game. She lost 43 points during the second game. There were 3428 people watching in the studio. Susana earned 4086 points, then 3 more points to win the game! She won a trip to 43 different countries. Add the points that Susana earned.

Lesson 60

Facts	+	Problems	+	Bonus	=	TOTAL

1

8 + 3	8 + 2	8 + 5	8 + 4	8 + 5	8 + 1	8 + 3	8 + 5
8 + 4	8 + 3	8 + 2	8 + 5	8 + 3	8 + 4	8 + 3	8 + 5

2

A 400 + 500 = ______

B 80 + 70 = ______

C 10 + 70 = ______

D 100 + 400 = ______

E 60 + 60 = ______

3

3 + 4	7 + 4	5 + 4	7 + 4	2 + 4	6 + 4	3 + 4	3 + 4
5 + 4	7 + 4	3 + 4	6 + 4	4 + 4	5 + 4	2 + 4	7 + 4

4

3 + 6	2 + 9	6 + 5	9 + 6	4 + 6	9 + 2	4 + 5	6 + 7
7 + 8	6 + 4	3 + 9	4 + 7	9 + 9	5 + 6	8 + 8	4 + 7

5

A 16 + 6 = ______

B 16 + 3 = ______

C 28 + 8 = ______

D 46 + 5 = ______

E 14 + 4 = ______

F 37 + 9 = ______

G 17 + 9 = ______

H 42 + 5 = ______

I 17 + 4 = ______

J 21 + 8 = ______

6

A ________

B ________

C ________

D ________

E ________

F ________

7

Add each number to 17.

8	2	5	3	9	0	7	6

8

A
$$\begin{array}{r} 1393 \\ 696 \\ 2482 \\ +\ \ 736 \\ \hline \end{array}$$

B
$$\begin{array}{r} 262 \\ 666 \\ 570 \\ +411 \\ \hline \end{array}$$

C
$$\begin{array}{r} 4169 \\ 377 \\ 1026 \\ +\ \ 145 \\ \hline \end{array}$$

D
$$\begin{array}{r} 1510 \\ 3660 \\ 1610 \\ +1210 \\ \hline \end{array}$$

A

Dr. Whitehead was an animal doctor. She treated 15 dogs and 32 cats in the morning. She gave 17 shots and 25 pills to animals. She treated 19 birds. How many animals did Dr. Whitehead treat?

B

The salesman went to 97 stores. He went to 4688 houses. He talked to 435 people. He went to 15 offices. He went to 974 restaurants. How many places did the salesman go to?

C

Allen cleaned out the garage for 9 hours. He talked on the phone for 10 hours. He cleaned out the attic for 17 hours. He cleaned the refrigerator for 12 hours. He watched television for 5 hours. He cut wood for 15 hours. How many hours did Allen work?

D

The machine made 945 chocolate bunnies. It made 38 chocolate drops. It made 156 coconut drops and 4 lemon drops. It made 96 chocolate bars. How many pieces of chocolate candy did the machine make?

E

Shoppers bought 243 radios. They bought 165 pairs of shoes. They looked at 24 lamps. They bought 57 suits. Count the things the shoppers will wear.

Lesson 61

Test	+	Facts	+	Problems	+	Bonus	=	TOTAL

1

8	8	8	8	8	8	8	8
+ 7	+ 10	+ 5	+ 6	+ 8	+ 7	+ 6	+ 8

8	8	8	8	8	8	8	8
+ 7	+ 5	+ 6	+ 8	+ 7	+ 6	+ 9	+ 8

2

A 900 + 500 = ________ **B** 30 + 70 = ________

C 50 + 70 = ________ **D** 400 + 400 = ________

E 20 + 70 = ________

3

8	8	8	8	8	8	8	8
+ 3	+ 1	+ 5	+ 2	+ 4	+ 5	+ 5	+ 3

8	8	8	8	8	8	8	8
+ 4	+ 2	+ 3	+ 5	+ 3	+ 4	+ 5	+ 2

4

A 12 + 4 = ______ **B** 19 + 6 = ______ **C** 12 + 5 = ______

D 17 + 8 = ______ **E** 15 + 9 = ______ **F** 12 + 7 = ______

G 15 + 8 = ______ **H** 13 + 6 = ______ **I** 14 + 4 = ______

J 13 + 7 = ______

5

9 + 9	2 + 6	6 + 6	3 + 9	2 + 6	5 + 9	6 + 3	4 + 7
7 + 9	3 + 6	3 + 3	4 + 7	5 + 6	9 + 8	6 + 7	7 + 7
8 + 7	6 + 4	8 + 8	4 + 4	2 + 9	9 + 4	4 + 6	4 + 5
9 + 2	5 + 8	9 + 6	2 + 8	4 + 6	5 + 5	7 + 9	2 + 9

6

A ________ **B** ________ **C** ________ **D** ________

E ________ **F** ________

7

A	**B**	**C**	**D**
214 984 699 + 177	132 374 159 + 154	2190 2591 3373 + 1454	3418 1079 + 4295

A

261 teachers went swimming. 493 students built forts. 97 parents went swimming. 135 students rode bikes. 2580 children went swimming. How many people went swimming?

B

Bernardo stacked 563 bricks. Lola stacked 3470 bricks. Lola stacked 72 bricks. Lola stacked 14 plates. Lola stacked 627 bricks. How many bricks did Lola stack?

C

83 jet airplanes flew over the city. 4 eagles flew over the city. 103 pigeons flew over the city. 792 bluebirds flew over the farm. 845 bluebirds flew over the city. How many birds flew over the city?

D

Shoji threw away 81 dog whistles. He found 128 dog whistles. He sold 12 dogs. He sold 41 dog whistles. He tossed 38 dog whistles in the garbage can. How many dog whistles did Shoji throw away?

E

Jill's mom gave her 84 silver coins. Jill lost 30 silver coins. Her father gave her 90 more silver coins. Jill found 5 silver coins and 8 gold coins in a box in the attic. How many silver coins did Jill get?

Lesson 62

Facts + Problems + Bonus = TOTAL

1

5	3	4	5	2	3	2	5
+ 8	+ 8	+ 8	+ 8	+ 8	+ 8	+ 8	+ 8

4	2	5	3	3	5	2	4
+ 8	+ 8	+ 8	+ 8	+ 8	+ 8	+ 8	+ 8

2

A 30 + 60 = ________

B 700 + 800 = ________

C 90 + 50 = ________

D 30 + 30 = ________

E 600 + 600 = ________

3

8	8	8	8	8	8	8	8
+ 9	+ 6	+ 7	+ 9	+ 8	+ 6	+ 8	+ 6

8	8	8	8	8	8	8	8
+ 8	+ 9	+ 6	+ 7	+ 6	+ 9	+ 6	+ 7

4

4	7	7	6	4	6	6	9
+ 5	+ 7	+ 4	+ 7	+ 6	+ 5	+ 4	+ 4

3	8	4	7	6	5	6	7
+ 9	+ 7	+ 7	+ 5	+ 6	+ 9	+ 5	+ 9

5

A $17 + 6 =$ ______ B $15 + 9 =$ ______ C $13 + 6 =$ ______

D $18 + 7 =$ ______ E $13 + 7 =$ ______ F $13 + 5 =$ ______

G $17 + 7 =$ ______ H $13 + 3 =$ ______ I $15 + 6 =$ ______

J $19 + 7 =$ ______

6

A ________ B ________ C ________ D ________

E ________ F ________

7

Add each number to 11.

8	3	6	9	2	4	5	7

8

A
```
 2185
  186
 1975
+1386
-----
```

B
```
 38
 18
 29
+47
---
```

C
```
 1910
 1860
 2870
+1930
-----
```

D
```
  779
  195
+  68
-----
```

A

The students read 104 stories. They wrote 186 stories. They read 425 poems. They told 73 stories. They wrote 81 stories. They wrote 377 stories. How many stories did the students write?

B

Madge ate 3 baked potatoes. She bought 7 potatoes. She ate 2 fried potatoes. She ate 2 boiled eggs. She ate 3 boiled potatoes. How many potatoes did Madge eat?

C

In 2000, the sailing team won 85 races and lost 32 races. In 2001, the sailing team won 94 races and lost 239 races. In 2002, the team won 88 races and lost 499 races. How many races did the sailing team lose?

D

The store gave away 30 birdhouses. It sold 14 old birdhouses. It bought 280 boxes of bird seed. It sold 2130 new birdhouses. It sold 150 orange birdhouses. How many birdhouses did the store sell?

Lesson 63

Facts	+	Problems	+	Bonus	=	TOTAL

1

8 + 6	8 + 10	8 + 7	8 + 8	8 + 6	8 + 7	8 + 6	8 + 9
8 + 6	8 + 8	8 + 9	8 + 5	8 + 7	8 + 6	8 + 9	8 + 6

2

A 50 + 70 = ________ B 30 + 60 = ________

C 500 + 200 = ________ D 800 + 800 = ________

E 90 + 90 = ________

3

5 + 8	3 + 8	2 + 8	4 + 8	5 + 8	2 + 8	4 + 8	3 + 8
5 + 8	4 + 8	3 + 8	2 + 8	5 + 8	3 + 8	5 + 8	4 + 8

4

4 + 6	9 + 9	7 + 9	7 + 4	6 + 3	9 + 6	9 + 5	6 + 7
6 + 4	7 + 5	4 + 7	4 + 9	5 + 6	7 + 6	3 + 6	6 + 5

A 19 + 5 = ______

B 16 + 6 = ______

C 14 + 4 = ______

D 17 + 4 = ______

E 17 + 7 = ______

F 15 + 8 = ______

G 13 + 6 = ______

H 12 + 9 = ______

I 12 + 8 = ______

J 14 + 3 = ______

6

A ________

B ________

C ________

D ________

E ________

F ________

7

Add each number to 14.

3	7	2	5	6	8	4	9

8

A

$$\begin{array}{r} 3481 \\ 8080 \\ 2192 \\ +\,5881 \\ \hline \end{array}$$

B

$$\begin{array}{r} 686 \\ 951 \\ 361 \\ +\,101 \\ \hline \end{array}$$

C

$$\begin{array}{r} 1818 \\ 1943 \\ 2775 \\ +\quad 559 \\ \hline \end{array}$$

D

$$\begin{array}{r} 3260 \\ 2670 \\ 1980 \\ +\quad 140 \\ \hline \end{array}$$

A

The repair shop fixed 2130 red cars. It fixed 150 blue trucks. It fixed 86 green cars. It fixed 210 black cars. How many cars did the repair shop fix?

B

Sola made 43 cups. She made 28 plates. Then she made 72 glasses. Sola made 53 more cups. How many cups did Sola make?

C

A family of 85 red ants lived underneath the porch steps. Another family of 1428 red ants lived underneath the apple tree. A third family of 2143 ants lived in the kitchen, behind the stove. 834 more red ants went to live under the apple tree. How many ants ended up living under the apple tree?

D

3428 people went to the ball game Monday. 948 people went to the park. 823 people went to the ball game Tuesday. 3105 went to the ball game Wednesday. How many people went to the ball game?

E

The factory made 342 big cars. It made 1432 small trucks. It made 6428 small cars. It made 84 sports cars. It gave away 25 cars. How many cars did the factory make?

Lesson 64

Facts	+	Problems	+	Bonus	=	TOTAL

1

$$\begin{array}{r}4\\+8\\\hline\end{array}\quad\begin{array}{r}2\\+8\\\hline\end{array}\quad\begin{array}{r}5\\+8\\\hline\end{array}\quad\begin{array}{r}1\\+8\\\hline\end{array}\quad\begin{array}{r}3\\+8\\\hline\end{array}\quad\begin{array}{r}5\\+8\\\hline\end{array}\quad\begin{array}{r}4\\+8\\\hline\end{array}\quad\begin{array}{r}3\\+8\\\hline\end{array}$$

$$\begin{array}{r}3\\+8\\\hline\end{array}\quad\begin{array}{r}2\\+8\\\hline\end{array}\quad\begin{array}{r}4\\+8\\\hline\end{array}\quad\begin{array}{r}5\\+8\\\hline\end{array}\quad\begin{array}{r}4\\+8\\\hline\end{array}\quad\begin{array}{r}3\\+8\\\hline\end{array}\quad\begin{array}{r}5\\+8\\\hline\end{array}\quad\begin{array}{r}2\\+8\\\hline\end{array}$$

2

A $80 + 80 =$ ________

B $40 + 20 =$ ________

C $700 + 700 =$ ________

D $90 + 10 =$ ________

E $30 + 30 =$ ________

3

A ☐ { 8 ____________________ ; 6 ____________________ }

B ☐ { 8 ____________________ ; 10 ____________________ }

C ☐ { 8 ____________________ ; 9 ____________________ }

D ☐ { 8 ____________________ ; 7 ____________________ }

E ☐ { 8 ____________________ ; 6 ____________________ }

F ☐ { 8 ____________________ ; 8 ____________________ }

4

4	6	7	6	4	7	3	9
+ 7	+ 6	+ 6	+ 4	+ 9	+ 9	+ 9	+ 9

4	9	5	3	8	7	6	9
+ 6	+ 6	+ 6	+ 6	+ 7	+ 4	+ 7	+ 5

5

A 15 + 7 = ______ **B** 13 + 9 = ______ **C** 18 + 5 = ______

D 13 + 4 = ______ **E** 12 + 8 = ______ **F** 14 + 3 = ______

G 15 + 3 = ______ **H** 17 + 9 = ______ **I** 12 + 7 = ______

J 14 + 8 = ______

6

A ________ **B** ________ **C** ________ **D** ________

E ________ **F** ________

7

Add each number to 19.

7	2	8	3	6	4	9	5
☐	☐	☐	☐	☐	☐	☐	☐

8

A	B	C	D
458	883	218	864
186	685	433	86
124	351	95	208
+ 231	+ 91	+ 137	+ 124

A

The dancer learned 4 dance steps. He learned 2108 dance steps. He danced in 158 shows. He had 83 pairs of dancing shoes. He learned 4125 more dance steps. How many dance steps did the dancer learn?

B

28 men watched the people fishing. 36 men went fishing. 142 girls went fishing. 36 women went fishing. 8 dogs went fishing. How many people went fishing?

C

The man was a model. He modeled 92 coats. He modeled 85 hats. He owned 27 ties. He modeled 314 jackets. He changed clothes 341 times. How many things did the man model?

D

We found 824 white mice. We bought 135 white mice. We sold 942 white mice. We gave away 81 white mice. Our friend gave us 8 white mice. How many white mice did we get?

E

We found 342 food trays in the school lunchroom. We found 25 empty milk boxes in the lunchroom. We picked up 148 trays and 37 orange peels in the hall. Finally, we found 2413 trays on the picnic grounds. How many trays did we find in all?

Lesson 65

Test	+	Facts	+	Problems	+	Bonus	=	TOTAL

1

6	9	7	9	7	6	5	7
+ 8	+ 8	+ 8	+ 8	+ 8	+ 8	+ 8	+ 8

6	9	5	8	6	7	5	8
+ 8	+ 8	+ 8	+ 8	+ 8	+ 8	+ 8	+ 8

2

A 80 + 20 = ________ **B** 50 + 20 = ________

C 700 + 700 = ________ **D** 30 + 60 = ________

E 90 + 40 = ________

3

2	5	3	4	5	3	3	5
+ 8	+ 8	+ 8	+ 8	+ 8	+ 8	+ 8	+ 8

3	5	4	2	2	5	4	3
+ 8	+ 8	+ 8	+ 8	+ 8	+ 8	+ 8	+ 8

4

2	9	3	9	3	7	3	8
+ 9	+ 6	+ 7	+ 5	+ 8	+ 6	+ 4	+ 7

5	9	3	6	8	5	4	7
+ 5	+ 8	+ 6	+ 9	+ 3	+ 7	+ 9	+ 5

Part 4 continues on the next page.

7	8	8	7	4	8	6	6
+ 8	+ 9	+ 5	+ 7	+ 6	+ 8	+ 5	+ 8

4	6	3	4	4	2	7	6
+ 7	+ 6	+ 9	+ 5	+ 8	+ 6	+ 9	+ 4

9	7	3	6	2	2	8	9
+ 7	+ 8	+ 5	+ 7	+ 7	+ 5	+ 6	+ 4

8	5	5	5	4	9	5	7
+ 4	+ 8	+ 6	+ 9	+ 4	+ 9	+ 4	+ 3

5

A ________ B ________ C ________ D ________

E ________ F ________

6

Add each number to 19.

4	8	5	7	9	2	6	1

7

A	B	C	D
1326	1090	3829	3895
8489	189	788	1486
486	372	866	802
+ 848	+ 168	+ 659	+ 1976

8

A

94 cats went to the park. 37 children went to the park. 214 men went to the movie theater. 158 women went to the park. 342 men went to the park. How many people went to the park?

B

Ramon caught 24 grasshoppers in his backyard. His sister caught 348 grasshoppers in the yard. Ramon watched 3184 grasshoppers. He also saw 8 spiders. He found 6 grasshoppers sitting on a picnic table. How many grasshoppers did Ramon see?

C

The family cleaned out its attic and had a yard sale. It sold 350 books. It gave away 80 balls. It sold 250 sets of cards. It threw away 340 broken toys. It sold 435 old hats. How many things did the family sell?

D

418 boys ran in the race. 27 boys watched the race. 140 girls ran to the park. 620 women ran in the race. 342 girls ran in the race. How many people ran in the race?

Transition Lesson 8

BONUS

1

$$\begin{array}{r} 5 \\ +4 \\ \hline \end{array} \quad \begin{array}{r} 5 \\ +2 \\ \hline \end{array} \quad \begin{array}{r} 5 \\ +4 \\ \hline \end{array} \quad \begin{array}{r} 5 \\ +3 \\ \hline \end{array} \quad \begin{array}{r} 5 \\ +2 \\ \hline \end{array} \quad \begin{array}{r} 5 \\ +4 \\ \hline \end{array} \quad \begin{array}{r} 5 \\ +3 \\ \hline \end{array} \quad \begin{array}{r} 5 \\ +4 \\ \hline \end{array}$$

2

A 7 { 6 — $6 + 1 = 7$; 1 — $1 + 6 = 7$

B 4 { 3 — ________; 1 — ________

C 5 { 4 — ________; 1 — ________

D 10 { 9 — ________; 1 — ________

3

A 279 **B** 146 **C** 657 **D** 328 **E** 864

4

6	3	8	7	2	9	1	4
[ ]	[ ]	[ ]	[ ]	[ ]	[ ]	[ ]	[ ]

Transition Lesson 23

BONUS

1

10	10	10	10	10	10	10	10
+ 3	+ 1	+ 2	+ 5	+ 4	+ 3	+ 2	+ 5

2

A 7: 6, 1
6 + 1 = 7
1 + 6 = 7

B 4: 3, 1

C 5: 4, 1

D 10: 9, 1

3

A 279 **B** 146 **C** 657 **D** 328 **E** 864

4

A Add the plants.

+

B Add the CDs.

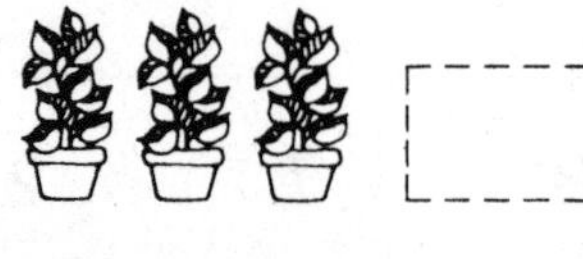

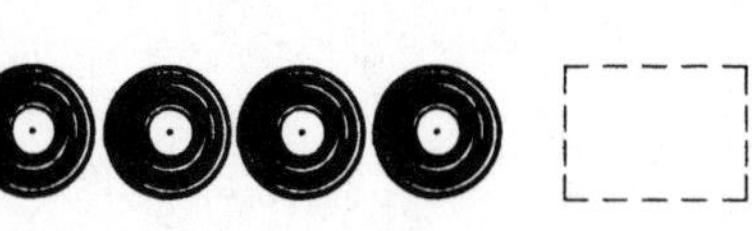

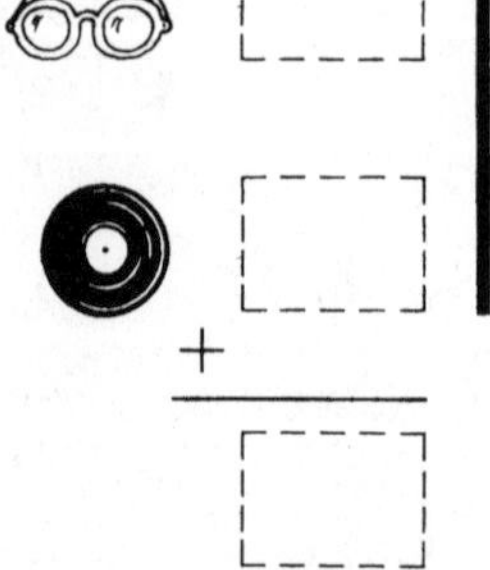

+

C Add the clocks.

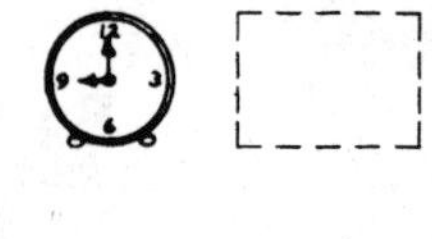

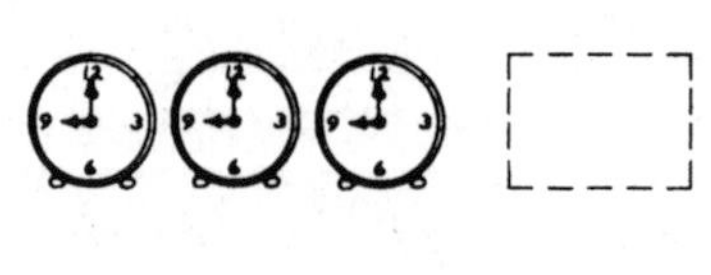

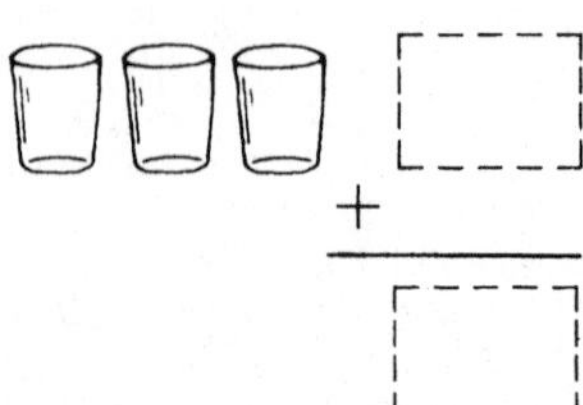

+

Transition Lesson 23 (continued)

5

A

Add the number of shirts Sam sews.

1. Sam sews 5 shirts. ___
2. Sam sells 7 shirts. ___
3. Sam sews 3 jackets. ___
4. Sam sews 4 shirts. ___
5. Sam sews 1 shirt. + ___

[ ] shirts

B

Add the number of eggs Carol eats.

1. Carol eats 1 egg. ___
2. Carol cooks 1 potato. ___
3. Carol eats 4 eggs. ___
4. Carol eats 2 eggs. ___
5. Carol cooks 3 eggs. + ___

[ ] eggs

6

A	B	C	D	E	F
21	30	12	14	13	10
33	23	12	31	41	13
31	11	32	10	31	32
+ 12	+ 12	+ 31	+ 21	+ 11	+ 20

Mastery Test Review—Lesson 7

1

A 406 B 308 C 728 D 507 E 346 F 205

2

A 508 B 326 C 409 D 583 E 704 F 542

Mastery Test Review—Lesson 11

1

A
7 { 6 ______ / 1 ______

B
4 { 3 ______ / 1 ______

C
5 { 4 ______ / 1 ______

D
10 { 9 ______ / 1 ______

Mastery Test Review Lesson 11

2

A

6 { 6 ______ 0 ______

B

4 { 3 ______ 1 ______

C

11 { 10 ______ 1 ______

D

3 { 3 ______ 0 ______

E

4 { 4 ______ 0 ______

F

8 { 7 ______ 1 ______

Mastery Test Review—Lesson 16

1

A 425 **B** 738 **C** 916 **D** 437 **E** 514

2

A 305 + 21 = ______

B 420 + 8 = ______

C 350 + 104 = ______

Mastery Test Review—Lesson 21

1

2	3	1	4	0	5	2	4
+ 5	+ 5	+ 5	+ 5	+ 5	+ 5	+ 5	+ 5

3	5	1	4	0	3	2	5
+ 5	+ 5	+ 5	+ 5	+ 5	+ 5	+ 5	+ 5

2

2	2	2	2	2	2	2	2
+ 4	+ 1	+ 3	+ 2	+ 5	+ 3	+ 0	+ 4

2	2	2	2	2	2	2	2
+ 3	+ 5	+ 4	+ 2	+ 4	+ 5	+ 3	+ 2

Mastery Test Review—Lesson 25

1

A	B	C	D	E	F
51	52	24	11	35	41
52	53	22	41	21	52
21	11	50	53	10	15
+ 12	+ 11	+ 11	+ 12	+ 2	+ 1
6	7				

2

A	B	C	D	E	F
41	84	52	52	91	13
12	22	42	1	11	42
55	31	11	+ 15	10	42
+ 41	+ 21	+ 10		+ 12	+ 1
9					

Mastery Test Review—Lesson 31

1

$2 + 10$ | $1 + 10$ | $5 + 10$ | $8 + 10$ | $3 + 10$ | $9 + 10$ | $6 + 10$ | $1 + 10$

$7 + 10$ | $4 + 10$ | $6 + 10$ | $2 + 10$ | $5 + 10$ | $8 + 10$ | $3 + 10$ | $2 + 10$

2

$5 + 7$ | $5 + 9$ | $5 + 8$ | $5 + 6$ | $5 + 9$ | $5 + 7$ | $5 + 9$ | $5 + 6$

$5 + 6$ | $5 + 7$ | $5 + 9$ | $5 + 6$ | $5 + 8$ | $5 + 5$ | $5 + 7$ | $5 + 9$

Mastery Test Review—Lesson 35

1

A ☐ $35 + 59 + 11$

B ☐ $11 + 29 + 49 + 11$

C ☐ $45 + 25 + 54$

D ☐ $81 + 14 + 25$

2

A $31 + 19 + 22$

B $21 + 24 + 55 + 43$

C $75 + 25 + 15 + 12$

D $33 + 122 + 415$

E $51 + 19 + 24 + 11$

Mastery Test Review—Lesson 41

1

8	6	7	9	8	6	7	9
+ 5	+ 5	+ 5	+ 5	+ 5	+ 5	+ 5	+ 5

5	9	6	8	6	9	7	8
+ 5	+ 5	+ 5	+ 5	+ 5	+ 5	+ 5	+ 5

2

4	2	5	1	3	5	4	3
+ 7	+ 7	+ 7	+ 7	+ 7	+ 7	+ 7	+ 7

5	3	5	4	2	5	3	4
+ 7	+ 7	+ 7	+ 7	+ 7	+ 7	+ 7	+ 7

Mastery Test Review—Lesson 42

A

Susan bought 4 cakes. She gave away 2 cakes. Her friend gave her 5 cakes. She found 1 cake. She bought 2 pies. Add the cakes that Susan got.

B

Julia played cards for 2 hours. She played ball for 3 hours. She did homework for 2 hours. Julia jumped rope for 2 hours. Add the hours that Julia played.

Part 1 continues on the next page.

C

Hiro whistled 20 times. He sang 40 times. He whistled 30 times. He shouted 10 times. He whistled 50 times. Add the times Hiro whistled.

D

Rosa bought 25 rocks for her collection. Trudy bought 14 rocks. Rosa found 30 rocks. She gave away 10 rocks. She found 8 bricks. How many rocks did Rosa get?

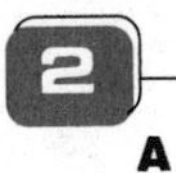

A

Leroy borrowed 4 eggs. He scrambled 1 egg for lunch. He bought 6 eggs. Someone gave him 1 apple. He bought 6 eggs. Add the eggs Leroy got.

B

The tractor dug rows for 21 hours. It cut grass for 32 hours. It loaded corn for 22 hours. It sat in the field for 41 hours. It planted seeds for 13 hours. How many hours did the tractor work?

C

Maya fed 5 horses. She rode 3 horses. She fed 5 pigs. She fed 6 cats. She ran after 8 chickens. Add the animals that Maya fed.

D

Nancy made 43 shirts. She washed 25 shirts. She made 12 hats. She made 53 jackets. She made 3 sandwiches. Add the pieces of clothing that Nancy made.

Mastery Test Review—Lesson 46

1

A
```
  351
  154
  242
+ 111
-----
```

B
```
  213
  192
  393
+ 111
-----
```

C
```
  971
  453
  232
+  41
-----
```

D
```
  394
  111
  323
+ 111
-----
```

E
```
  671
  132
   21
+ 135
-----
```

F
```
  471
  543
  222
+ 211
-----
```

Mastery Test Review—Lesson 47

1

A 4006 **B** 5027 **C** 7004 **D** 2902

E 4098 **F** 7007

2

A 5009 **B** 9035 **C** 8024 **D** 7406

E 9004 **F** 8201 **G** 5034

Mastery Test Review—Lesson 51

1

A	B	C	D	E	F
5	2	9	7	4	6
+ 9	+ 9	+ 9	+ 9	+ 9	+ 9

2

6	5	3	4	7	4	5	7
+ 6	+ 5	+ 3	+ 4	+ 7	+ 4	+ 5	+ 7
7	6	2	7	6	4	3	6
+ 7	+ 6	+ 2	+ 7	+ 6	+ 4	+ 3	+ 6

3

7	7	7	7	7	7	7	7
+ 9	+ 8	+ 7	+ 6	+ 9	+ 5	+ 8	+ 6
7	7	7	7	7	7	7	7
+ 7	+ 6	+ 8	+ 6	+ 5	+ 9	+ 6	+ 8

Mastery Test Review—Lesson 55

1

A ________ B ________ C ________ D ________

2

A ________ B ________ C ________ D ________

Mastery Test Review—Lesson 61

1

9 + 6	9 + 4	9 + 9	9 + 7	9 + 5	9 + 9	9 + 8	9 + 7
9 + 2	9 + 7	9 + 3	9 + 6	9 + 2	9 + 6	9 + 3	9 + 8

2

3 + 6	5 + 6	4 + 6	2 + 6	3 + 6	5 + 6	6 + 6	4 + 6
5 + 6	6 + 6	5 + 6	3 + 6	4 + 6	4 + 6	3 + 6	5 + 6

Mastery Test Review Lesson 61

3

4	9	4	5	8	9	4	6
+ 3	+ 7	+ 5	+ 7	+ 7	+ 6	+ 6	+ 7

9	4	2	4	8	4	6	9
+ 9	+ 7	+ 9	+ 6	+ 8	+ 9	+ 6	+ 3